PROGRAMMING MATHEMATICS USING MATLAB®

PROGRAMMING MATHEMATICS USING MATLAB®

LISA A. OBERBROECKLING
Department of Mathematics and Statistics
Loyola University Maryland
Baltimore, MD, United States

Academic Press is an imprint of Elsevier
125 London Wall, London EC2Y 5AS, United Kingdom
525 B Street, Suite 1650, San Diego, CA 92101, United States
50 Hampshire Street, 5th Floor, Cambridge, MA 02139, United States
The Boulevard, Langford Lane, Kidlington, Oxford OX5 1GB, United Kingdom

Library of Congress Cataloging-in-Publication Data
A catalog record for this book is available from the Library of Congress

British Library Cataloguing-in-Publication Data
A catalogue record for this book is available from the British Library

ISBN: 978-0-12-817799-0

For information on all Academic Press publications
visit our website at https://www.elsevier.com/books-and-journals

Publisher: Katey Birtcher
Editorial Project Manager: Rafael G. Trombaco
Production Project Manager: Beula Christopher
Designer: Bridget Hoette

Typeset by VTeX

To Rob and James, for the support, patience (especially when teaching MATLAB), and laughter.

To Christos Xenophontos for your encouragement and for introducing me to MATLAB.

Contents

Preface

This book started in 2004 when I started to use MATLAB® in my courses at Loyola University Maryland. I started including a few MATLAB projects within the introduction to linear algebra and multivariable calculus courses. I also taught a one-credit MATLAB course that was required of every mathematics major. Later the course was changed to a three-credit course. I expanded the previous assignments and added some others. The class notes and assignments from over the years have expanded into this book.

My philosophy has always been to use MATLAB to practice basic programming skills with mathematics topics students had seen previously, such as numerical integration, and topics they likely had not seen such as fractals. Visualizing mathematics has always been important.

Supplements

Student companion site: Please visit https://www.elsevier.com/books-and-journals/book-companion/9780128177990

Instructor-only site: Qualified instructors can register and access teaching materials at https://textbooks.elsevier.com/web/Manuals.aspx?isbn=9780128177990

Introduction

The goal of the course and thus book is to introduce MATLAB® and to practice basic programming techniques. There is a lot more to MATLAB than what is covered in this book. Most students have already had some programming experience in another language before taking the course this book has stemmed from, although it is not necessary. On completion of the course/book, one should be familiar enough with MATLAB to explore more complicated features and commands. Deepening your understanding of mathematics and learning new topics are bonuses!

PART 1

MATLAB®

CHAPTER 1

Introduction to MATLAB®

1.1. Basic MATLAB® information

1.1.1 Starting MATLAB

MATLAB has many different windows or panels, the first three of which are on the main screen by default (see Table 1.1).

You can always modify the layout of the panels including "docking" or "undocking" them, in the "Home" view, select "Layout" from the menu and the top item is "Default." In order to use MATLAB successfully, you should pay attention to the `Current Folder`. Otherwise, MATLAB may not be able to save and run your programs/files successfully.

1.1.2 Good commands to know

The first four commands are useful to "start fresh" without closing and reloading MATLAB.

- `clc` Clears the command window of all previous commands and output. These commands are still stored in the command history and can be accessed with the up-arrow.
- `clear` Clears all defined variables in memory. BE CAREFUL WITH THIS COMMAND! You can also clear certain variable names by typing `clear varname1 varname2`.
- `clf` and `close` Clears the current figure (plot). If no figure window was open, it will open a blank figure window. Another command is `close`. This will close the current figure window. The command `close all` will close all figure windows. Note that there is no `clf all`. Both of these commands have other variations to

Table 1.1 Main MATLAB Windows.

Command Window	Enter commands and variables, run programs
Workspace Window	Information about the current variables
Current Folder Window	Shows the files in the current folder/directory
Command History Window	History of commands entered in Command Window
Figure Window	Output from graphic commands
Editor Window	Creates and debugs script and function files
Help Window	Help information
Launch Pad Window	Access to tools, demos, and documentation

Programming Mathematics Using MATLAB®
https://doi.org/10.1016/B978-0-12-817799-0.00006-5

clear/close named figures, etc. such as `clf(2)` and `close(2)` that will clear or close Figure 2, respectively.

- `format` Sends the output display back to the default format. The command `format compact` will not have as much white space (blank lines) between subsequent commands and output.
- `exit` or `quit` Quit MATLAB. You can also quit/exit MATLAB by using the MENU option under "File" or the usual close application icon within a Mac or Windows environment.

Other good commands

- `who` Lists current variables
- `help command` Displays the help for `command`. For example, `help atan2`.
- `disp('text')` Displays `text` as output in the command window.
- `exist text` or `exist('text')` Checks if variables or functions are defined (see `help exist` for more details).
- `lookfor text` Searches for "text" as a keyword in help entries of functions.

```
>> exist average

ans =

     0

>> lookfor average
mean                            - Average or mean value.
HueSaturationValueExample       - Compute Maximum Average HSV of Images with MapReduce
ewmaplot                        - Exponentially weighted moving average chart.
```

1.2. Basic mathematics

Mathematical operations

MATLAB uses the typical symbols for addition, subtraction, multiplication, division, and exponentiation (+, -, *, /, and ^). These are considered **matrix arithmetic operations** and follow the rules from linear or matrix algebra.

```
>> 10/3

ans =

    3.3333
```

MATLAB has another division operator, the "divided into" operator.

```
>> 10\3

ans =

    0.3000
```

Thus the forward slash is our usual division, "3 divided by 2" while the backslash is "3 divided into 2." The need for both of these become more apparent when working with matrices.

"Dot" operations or component-wise operations are useful and/or necessary for use with vectors and matrices (discussed in Section 2.4). These are considered "array arithmetic operations" and are carried out element or component-wise.

MATLAB does NUMERICAL, rather than algebraic computations, as seen below. Think about what is expected versus what is given as the answer to the subraction calculation.

```
>> asin(1/2)

ans =

    0.5236

>> pi/6 - asin(1/2)

ans =

  -1.1102e-16
```

1.2.1 Built-in mathematical functions

Table 1.2 has common mathematical functions in MATLAB. This is not a complete list.

The **modulo** function `mod` calculates the modulus after division. In other words, `mod(x,y)` returns the remainder when you divide `x` by `y`. In some languages like Python or Perl this is equivalent to `%` is the modulus operator. Thus `mod(x,y)` in MATLAB is the same as `x % y` in other languages. There is also a **remainder** function `rem` that has the same functionality as `mod` EXCEPT when the divisor and quotient are opposite signs or the quotient is 0 (see Exercise 9).

```
>> mod(10,3)

ans =

     1
```

```
>> rem(10,3)

ans =

     1

>> mod(10,-3)

ans =

    -2

>> rem(10,-3)

ans =

     1

>> mod(10,0)

ans =

    10

>> rem(10,0)

ans =

   NaN
```

NOTE: `NaN` stands for "not a number."

1.2.2 Precedence rules

While some languages go strictly from left to right when there are multiple operations within one line, most now follow the mathematical rules for order of operations. Many of you may know the pneumonic PEMDAS (Please excuse my dear Aunt Sally); parentheses, exponentiation, multiplication, division, addition, subtraction. There are some issues with that pneumonic, however. There are also some discrepancies, to be explored in the exercises. Table 1.3 shows the precedence rules for the arithmetic operations discussed so far, going from highest precedence to lowest.

One thing to consider is: where do functions go on this list? For example, how does MATLAB interpret the command `cos(x)^2`? Is this equivalent to $\cos(x^2)$, $\cos^2(x)$, or something else entirely? This will be explored in the exercises.

Table 1.2 Mathematical functions.

sqrt(x)	Square root
exp(x)	Exponential (e^x)
abs(x)	Absolute value
mod(x,y)	Modulus
rem(x,y)	Remainder
log(x)	Natural logarithm ($\ln(x)$)
log10(x)	Common logarithm
sin(x)	Sine of x (radians)
sind(x)	Sine of x (degrees)
cos(x)	Cosine of x
tan(x)	Tangent of x
cot(x)	Cotangent of x
asin(x)	Inverse sine of x (radians)
asind(x)	Inverse sine of x (degrees)
pi	π

Table 1.3 Basic math precedence.

( )	parentheses
^	exponentiation
* / \	multiplication and division
+ -	addition and subtraction

If you have a long calculation/expression, you can continue on the next line in the command window or within a MATLAB file with the **ellipses** or **continuation operator**. Note when the ellipses work with/without spaces:

```
>> 1+2*3-12^2/3 ...
*2

ans =

   -89

>> 1+2*3-12^2/3 *...
2

ans =

   -89

>> 1+2*3-12^2/3*...
2
```

```
ans =

   -89

>> 1+2*3-12^2/3...
 1+2*3-12^2/3...
               |
Error: Unexpected MATLAB operator.
```

1.2.3 Formats

As mentioned above, the `format` command returns the format back to the default format, which is the same as `format short`. Generally speaking, this will display a number up to four decimal places, while `format long` will display 15. In scientific notation, this amounts to five and 16 significant digits, respectively. See Table 1.4.

```
>> pi

ans =

    3.1416

>> format long
>> pi

ans =

   3.141592653589793
```

There are other built-in formats, including how numbers in scientific notation are displayed. See `help format` for more examples.

Another useful command is `format compact` and `format loose` (default). This will change how the output is displayed.

```
>> pi

ans =

    3.1416

>> format compact
>> pi
ans =
    3.1416
>> format loose
```

```
>> pi

ans =

    3.1416

>>
```

The subsequent commands shown in this text will use `format compact` (see Table 1.4).

Table 1.4 Basic formats displaying `10*pi`.

`format short` (default)	`31.4159`
`format long`	`31.415926535897931`
`format rat`	`3550/113`
`format bank`	`31.42`
`format short e`	`3.1416e+01`
`format long e`	`3.141592653589793e+01`
`format short g`	`31.416`
`format long g`	`31.4159265358979`
`format hex`	`403f6a7a2955385e`
`format compact`	(no blank lines)
`format loose` (default)	(some blank lines)

1.3. Variables

The format for a variable assignment is as follows:

Variable name = Numerical value or computable expression

Some conventions:

- The = is the **assignment operator** which assigns a value to a variable.
- Left-hand side can include only **one** variable name.
- Right-hand side can be a number or an expression made up of numbers, functions, and/or variables previously assigned numerical values.
- Variables must begin with a letter.
- Press the **Enter/Return** key to make the assignment.
- The variable `ans` is the value of the last expression that is not assigned.
- Be careful with variable names. For example, do not name a variable `help` or `sin`.
- Variable names are case sensitive; thus `A` is not the same as `a`.

Remember:

- **Use semicolon (;) to suppress screen output**.
- Multiple commands can be typed on one line by typing a comma (,) between them if they are not already ended with a semicolon (;).

Example 1.3.1. Assign the number 3 to variable **a** and 4 to variable **b**. Print $\sqrt{a^2+b^2}$ and assign the solution to the variable **c**.

```
>> a=3; b=4; c = sqrt(a^2+b^2), a+b+c
c =
     5
ans =
    12
```

Notice in the above example, you do not need spaces around the "=" for variable assignments but you may use them for aesthetic reasons.

Example 1.3.2. Experiment with the equation

$$\cos^2\frac{x}{2}=\frac{\tan x+\sin x}{2\tan x}$$

by calculating each side of the equation for $x=\pi/5$.

```
>> x = pi/5;
>> LHS = (cos(x/2))^2, RHS = (tan(x)+sin(x))/(2*tan(x))
LHS =
    0.9045
RHS =
    0.9045
>> format long
>> LHS, RHS
LHS =
   0.904508497187474
RHS =
   0.904508497187474
>> LHS-RHS
ans =
    -1.110223024625157e-16
```

1.4. Diaries and script files

You can record your commands and output to the command window with the `diary` command. The commands you enter in the command window and any output are stored as an ASCII (plain text) file. The command `diary` toggles the recording on and off. If you do not specify a filename, it will create a file in the current folder of the name "diary." The command `diary filename` will save the recording to a file of the name "filename." The commands `diary off` and `diary on` will pause and restart the recording, respectively, to the active file. NOTE: when you use the `diary filename` more than once (within the same current folder), it will continue to APPEND to the file.

```
>> diary filename
>> 1+1
ans =
     2
>> diary off
>> 2+2
ans =
     4
>> diary on
>> 3^2
ans =
     9
>> diary
```

For example, the commands above will generate the following text in the file "filename":

```
1+1
ans =
     2
diary off
3^2
ans =
     9
diary
```

Script files, or m-files, are extremely useful for running and rerunning code. You may be required to turn in script files for your assignments.

- Script files are ASCII files (plain text files) with extension `.m`; thus they are also called m-files. These are basically batch files.
- When you run a SCRIPT file, MATLAB executes each line as if you typed each line into the command window.
- Script files are very useful; you can edit them, save them, execute them many times and "tweak" them to experiment with commands.
- The MATLAB editor window is the best way to create and edit script files.
- To avoid extraneous output to the command window, put ";" after variable assignments or intermediate calculations.
- Comments within MATLAB files begin with the percent symbol (`%`).

Running script files:

There are many ways to run an m-file of name `filename.m`. First, you must MAKE SURE CURRENT FOLDER IS CORRECT!

1. `>> filename`
2. `>> run filename`
3. `>> run('filename')`
4. Within the Editor tab, chose run...

1.5. Exercises

1. **Basic calculations** Use MATLAB to do the following calculations. Be careful! The following are displayed using regular mathematical notation; you need to figure out what MATLAB functions are needed.
   - **(a)** $\frac{7}{16}(2.4)(6^4) + \frac{.75^3}{2^8 - 225}$,
   - **(b)** $\frac{29^2}{5} + \frac{64^{4/3}}{11} + 20 \cdot 9^{-3}$,
   - **(c)** $\cos(360)$,
   - **(d)** $\cos(360°)$,
   - **(e)** $\cos(2\pi)$,
   - **(f)** $\cos(2\pi°)$,
   - **(g)** $e + 5$,
   - **(h)** $|\pi - 5|$,
   - **(i)** $3 \ln 7$,
   - **(j)** $3 \log 7$,
   - **(k)** $\frac{6}{\pi} \sin^{-1}(0.5) + 4$,
   - **(l)** $4 \cos(5 \arctan(13/4))$.
2. **Order of operations**
   - **(a)** Calculate, without using any parentheses, -4^2 using THREE of the following and write your answers on your own paper, specifying which you did and what answers they gave:
     - **(i)** calculator (specify type/model)
     - **(ii)** Google.com
     - **(iii)** Excel or Google Sheets
     - **(iv)** Desmos.com or Wolfram Alpha

     Now calculate -4^2 using MATLAB. Are there differences in the answers? Based on your knowledge of Order of Operations, what should be the answers?
   - **(b)** Do the same for the calculation of $-\cos(\pi/4)^2$ (you may use parentheses around the $\pi/4$; i.e. you should calculate `-cos(pi/4)^2`). NOTE: In Excel, to calculate with π use "PI()", again noting the differences (if any) in the answers.
   - **(c)** Do the same for the calculation of -2^{12} and $-8^{1/3}$ (without using any parentheses). Should parentheses be used to get the proper calculations? Where?
   - **(d)** How should Table 1.3 for precedence rules within MATLAB be changed or expanded to include functions and negation (unary minus)? Answer this by rewriting the table on paper.

3. **Using variables.** Define variables with the assignments $x = 8$, $y = 3.5$, and $X = 1/9$. Calculate the following within MATLAB. For the $\sqrt[n]{z}$ calculations, use the `nthroot` function.
   (a) $\dfrac{4(y-x)}{3X-20}$, **(b)** $\dfrac{3\sqrt{X}}{10}$,
   (c) $3\cos x \tan y$, **(d)** $e^{(X+y)/x} + 3^{\sqrt[3]{x}}$.
4. Suppose $x = 3$ and $y = 5$ (define the variables at the beginning of the problem).
   (a) $3\pi x^2$, **(b)** $\left(1 - \dfrac{1}{x^5}\right)^{-1}$, **(c)** $\dfrac{3y}{4x-8}$, **(d)** $\dfrac{4(y-5)}{3x-6}$, **(e)** $2\sin(x)\cos(y)$.
5. Define the following variables: `tablePrice` $= 1256.95$, `chairPrice` $= 89.99$, and `gasPrice` $= 3.499$. Using the variables and `format bank` for parts a–c, write your answers to the following questions on paper, interpreting the MATLAB output as a meaningful answer:
   (a) Find the cost of one table and eight chairs.
   (b) Find the same cost as above but with 6.5% sales tax.
   (c) Find the cost of 14.25 gallons of gas that you would have to pay.
   (d) Compute the *actual* cost of 14.25 gallons of gas using the **default format**.
6. **More calculations** Define the variables $x = 256$ and $y = 125$. Calculate the following within MATLAB. When radical notation is used in the problem, use the `sqrt` and `nthroot` functions and use exponential calculations when exponential notation is used in the problem.
   (a) $\sqrt{x}$, **(b)** $x^{\frac{1}{2}}$, **(c)** $\sqrt{-x}$, **(d)** $(-x)^{\frac{1}{2}}$, **(e)** $x^{\frac{1}{4}}$,
   (f) $\sqrt[4]{x}$, **(g)** $y^{\frac{1}{3}}$, **(h)** $-y^{\frac{1}{3}}$, **(i)** $(-y)^{\frac{1}{3}}$, **(j)** $\sqrt[3]{-y}$.
   (k) From the above calculations, do you see anything surprising with the answers?
   (l) Calculate $(-x)^{\frac{1}{4}}$ and $\sqrt[4]{-x}$. What are the differences?
   (m) Calculate $(-8)^{2/3}$ and $8^{2/3}$ on paper using your exponent rules. Now do the calculations within MATLAB, Excel, Google.com, WolframAlpha.com, and a scientific or graphing calculator (specifying what model you have used). What are the differences, if there are any?
7. **Calculator precision**
   (a) Within an Excel spreadsheet or Google Sheet, calculate $\sqrt[12]{1782^{12} + 1841^{12}}$ using exponential notation for the calculations. Write your answer clearly.
   (b) Rewrite the above expression with your answer to part (a) into an equation and algebraically simplify the equation so there are no radicals or rational exponents.
   (c) Now calculate $\sqrt[12]{1782^{12} + 1841^{12}}$ using MATLAB.
   (d) Compare the left-hand side and right-hand side of the equation you get in part (b) by subtraction within MATLAB, using `format long`).

(e) Calculate "3 quadrillion and 18 minus 3 quadrillion and 14" in your head and write it down. Now translate this into mathematics so you can calculate it within Google.com, Excel, and MATLAB. (You may need to look up how many zeros you will need!) Compare your answers in a table.

(f) Do the same with 2.000000000000018 – 2.000000000000014.

8. **Ambiguities with notation.** Define variables with the assignments $x = 10$ and $y = \pi/4$. Calculate the following within MATLAB. You may have to adjust from mathematical notation to correct MATLAB notation. Make sure you are using the default format!

(a) $\cos y$, **(b)** $\cos y^2$, **(c)** $\cos(y^2)$, **(d)** $\cos^2 y$, **(e)** $(\cos y)^2$,

(f) x^{-1}, **(g)** $\cos^{-1}(x/20)$, **(h)** $\cos(x/20)^{-1}$, **(i)** $(\cos(x/20))^{-1}$.

(j) Redo part (g) and then use the MATLAB variable `ans` to calculate

$$\frac{\cos^{-1}(x/20)}{4y}.$$

(k) Are any of the above calculations ambiguous in how they are written (which ones and why)? What could be done to make the calculations clearer to the person performing/entering the calculations?

9. **Exploring `rem` and `div`.** It is common to use either of the functions `mod` or `rem` to tell whether positive integers are even or odd, among other uses. We will explore the differences and similarities of these functions.

(a) Here is another simple use for these functions. You are given a list of 10-digit numbers. You would like to only use the last seven digits of these numbers (for example, for display purposes). Use both the `mod` or `rem` functions to easily get the last seven digits of the number `4108675309`. Do you see a difference in their use for this?

(b) Use the same commands on the number `-4108675309`. Do you see a difference? Explain in **your own words** what you think is the difference between the `mod` and `rem` functions. Is there a difference when using these functions to tell whether *any* integer is even or odd?

(c) Can you come up with a way, using MATLAB functions such as `mod`, `rem`, `round`, `ceil`, `fix`, etc. to capture the "area code" (first three digits) of 4108675309? Experiment with at least two phone numbers with different area codes.

(d) What about capturing the "central office" part of the number (867)? Do it for 4108675309 and 4106172000.

CHAPTER 2

Vectors and Matrices (Arrays)

2.1. One-dimensional arrays (vectors)

Arrays are used to store and manipulate numbers. They are arranged in rows and/or columns and are defined within MATLAB® using brackets `[ ]`.

One-dimensional arrays, usually called vectors, can represent points or vectors in space (of any dimension), or can be used to store data. For example: the point $(1, 2)$ in 2D or the point $(1, 2, -5)$ in 3D, or the data $(70, 75, 72, 77)$. These vectors can be represented as rows or columns.

To enter a row vector, use spaces or commas between numbers.

```
>> x = [1 2 3], y = [4, 5, 6]
x =
     1     2     3
y =
     4     5     6
```

To enter a column vector, use semi-colons between numbers:

```
>> y = [4;5;6]
y =
     4
     5
     6
```

There are ways to define vectors without having to enter every element. The two most common ways are to define constant spaced and equally spaced (linearly spaced) vectors.

2.1.1 Constant spaced vectors

If the **difference between the elements (increment)** of the vector is important, use "colon-notation". For a vector from m to n, incremented by q, use the format

$$\text{variablename} = [\text{m: q : n}]$$

If no increment `q` is given (as in the variable `x` below), it is 1 by default.

```
>> x=[1:7], y=[1:2:6], z=[1:3:25], w=[1:3:24]
x =
     1     2     3     4     5     6     7
y =
```


https://doi.org/10.1016/B978-0-12-817799-0.00007-7

```
     1     3     5
z =
     1     4     7    10    13    16    19    22    25
w =
     1     4     7    10    13    16    19    22
```

Notice in the vectors y and w above that the last number (n) may not actually be a value in the vector if the increment (q) is such that it will not occur based on the starting value (m).

One can also increment backwards with a negative increment value.

```
>> a=[6:-1:1], b=[10:-2:-2]
a =
     6     5     4     3     2     1
b =
    10     8     6     4     2     0    -2
```

2.1.2 Equally spaced vectors

If the **number of elements** in the vector is important but not necessarily the actual values, use `linspace`. The command `linspace` stands for linearly spaced. **This is especially useful for plotting.** The command

$$\text{variablename} = \text{linspace(m,n)}$$

will give you a vector of 100 equally-spaced elements between m and n. For a vector of q elements from m to n, use

$$\text{variablename} = \text{linspace(m,n,q)}$$

```
>> x = linspace(1,7,3), y = linspace(0,1,5)
x =
     1     4     7
y =
         0    0.2500    0.5000    0.7500    1.0000
```

Notice that both the values m and n are elements in the vector.

There is also a command `logspace` that is also useful for plotting (among other things). The command `logspace(a,b)` will generate a row vector of 50 logarithmically spaced points from 10^a and 10^b. If b is `pi`, then the points are between 10^a and π.

$$\text{variablename} = \text{logspace(a,b)}$$
$$\text{variablename} = \text{logspace(a,b,q)}$$

Just as in `linspace`, you can specify the number of elements in the vector by stating the value q.

```
>> logspace(0,5,6)
ans =
           1          10         100        1000       10000      100000
>> logspace(5,1,5)
ans =
      100000       10000        1000         100          10
>> linspace(1,pi,4)
ans =
    1.0000    1.7139    2.4277    3.1416
>> logspace(0,pi,4)
ans =
    1.0000    1.4646    2.1450    3.1416
>> logspace(2,pi,3)
ans =
  100.0000   17.7245    3.1416
```

2.2. Two-dimensional arrays (matrices)

Two-dimensional arrays or matrices can be used for many things. You can use matrices to store data or information as a table or spreadsheet. Matrices can also be used to solve systems of linear equations such as

$$
\begin{aligned}
2x + 3y + z &= 4,\\
x - 5y + 3z &= 3,\\
4x - 2y + 3z &= 2.
\end{aligned}
$$

The format for defining matrices expands on the format for defining row and column vectors; spaces or commas between elements within a row with semicolons between rows.

variable = [1st row ; 2nd row ; ... ; last row]

You can also use vectors of the same size to be rows (or columns) of a matrix, as we do to define the matrix `B` below.

```
>> A = [2 3 1;1 -5 3;4 -2 3; 0 1/2 66], x=0:4;...
y=linspace(0,pi,5); B=[x;y]
A =
    2.0000    3.0000    1.0000
    1.0000   -5.0000    3.0000
    4.0000   -2.0000    3.0000
         0    0.5000   66.0000
B =
         0    1.0000    2.0000    3.0000    4.0000
```

```
         0    0.7854    1.5708    2.3562    3.1416
```

2.3. Addressing elements of vectors/arrays

For many reasons, we may want to capture one element, or a portion of a vector or matrix. This section explains several ways to accomplish this.

Elements of vectors

- `v(k)` picks the kth element of `v`.
- `v(m:n)` picks the mth through the nth elements of `v`.

```
>> v = linspace(0,1,5)
v =
         0    0.2500    0.5000    0.7500    1.0000
>> v(3)
ans =
    0.5000
>> v(2:4)
ans =
    0.2500    0.5000    0.7500
```

Notice the difference if `v` is a matrix.

```
>> v=[1 2 3;4 5 6]
v =
     1     2     3
     4     5     6
>> v(2)
ans =
     4
>> v(3)
ans =
     2
>> v(3:6)
ans =
     2     5     3     6
```

Elements of matrices

- `A(m,n)` picks the (m, n)th element (element in the mth row, nth column) of the matrix `A`.
- `A(m:n, p:q)` gives the submatrix from the elements $(m : n) \times (p : q)$
- `A(m:n, :)` gives rows m through n, and every column of A

- `A(:, p:q)` gives every row, and columns p through q of A

```
>> A = [2 3 1;1 -5 3; 4 -2 3; 0 1 6]
A =
     2     3     1
     1    -5     3
     4    -2     3
     0     1     6
>> A(3,2)                % 3rd row, 2nd column of A
ans =
    -2
>> A(2:4,2:3)    %2-4 rows, 2-3 columns of A
ans =
    -5     3
    -2     3
     1     6
>> A(:,1)                %every row, 1st column of A
ans =
     2
     1
     4
     0
>> A(2:3,:)              %2-3 row, every column of A
ans =
     1    -5     3
     4    -2     3
```

Adding elements to arrays

You can easily add elements to arrays. You must be careful that what you are adding is of appropriate size.

```
>> B = [1 4 2 3;3 6 9 2;1 4 9 7]
B =
     1     4     2     3
     3     6     9     2
     1     4     9     7
>> C=[B; 2 5 1 8]              % adding a row to B
C =
     1     4     2     3
     3     6     9     2
     1     4     9     7
     2     5     1     8
>> D = [C [1;2;3;4]]     % adding a column to C
D =
     1     4     2     3     1
     3     6     9     2     2
```

```
      1     4     9     7     3
      2     5     1     8     4
```

Deleting elements

One can delete elements by assigning nothing to these elements.

```
>> v=[2:2:10]
v =
      2     4     6     8    10
>> v(2)=[]
v =
      2     6     8    10
>> v(2:3)=[]
v =
      2    10
>> B
B =
      1     4     2     3
      3     6     9     2
      1     4     9     7
>> B(2:3,:) = [ ]
B =
      1     4     2     3
>> C
C =
      1     4     2     3
      3     6     9     2
      1     4     9     7
      2     5     1     8

>> C(:,1:2) = [ ]
C =
      2     3
      9     2
      9     7
      1     8
```

Notice the colon: recall that this indicates "all columns" or "all rows."

Transpose

There are actually several commands for transposing an array, with subtle but important differences.

```
>> x, A=[1 2/3; -3 pi]
x =
```

```
      1      4      7
A =
    1.0000    0.6667
   -3.0000    3.1416
>> x', A'
ans =
      1
      4
      7
ans =
    1.0000   -3.0000
    0.6667    3.1416
>> x.', A.'
ans =
      1
      4
      7
ans =
    1.0000   -3.0000
    0.6667    3.1416
>> transpose(x), transpose(A)
ans =
      1
      4
      7
ans =
    1.0000   -3.0000
    0.6667    3.1416
```

```
>> format compact, format rat
>> B = [-1+2i, 2+3i;4-5i, 6-7i]
B =
       -1        +    2i             2        +    3i
        4        -    5i             6        -    7i
>> B'
ans =
       -1        -    2i             4        +    5i
        2        -    3i             6        +    7i
>> B.'
ans =
       -1        +    2i             4        -    5i
        2        +    3i             6        -    7i
>> transpose(B)
ans =
       -1        +    2i             4        -    5i
        2        +    3i             6        -    7i
>> ctranspose(B)
ans =
```

```
  -1 - 2i      4 + 5i
   2 - 3i      6 + 7i
```

Notice that if every number is real, there are no differences between the commands. But if there are complex numbers, there is a conjugate transpose, and a transpose (see if you can figure out which ones are which). Other good commands are `flip`, `fliplr`, `flipud`, etc.

Strings

Strings are arrays of characters rather than numbers, but addressing them can be done in a similar way.

- An array of characters
- Created by typing characters within single quotes
- Can include letters, digits, symbols and spaces

```
>> s = 'MATLAB is AWESOME'
s =
    'MATLAB is AWESOME'
>> s(5)
ans =
    'A'
>> s(1:6)
ans =
    'MATLAB'
>> s(1:6) = 'Maths*'
s =
    'Maths* is AWESOME'
```

2.4. Component-wise calculations

When doing calculations on arrays, one must be mindful as to whether the calculation is done component-wise or not and the sizes of what you are trying to combine. Just as in arithmetic on matrices, addition/subtraction and scalar multiplication is automatically done component-wise.

```
>> x=[1 2 3]; y=[4 5 6]; z=[10 11];
>> x+y
ans =
     5     7     9
>> x+z
Error using  +
Matrix dimensions must agree.
>> x+2
```

```
ans =
     3     4     5
>> -3*x
ans =
    -3    -6    -9
```

Multiplication of arrays is using the matrix multiplication definition.

```
>> x*y
Error using  *
Inner matrix dimensions must agree.
>> A=[1 2;3 4], B = [0 1;-1 1], A*B
A =
     1     2
     3     4
B =
     0     1
    -1     1
ans =
    -2     3
    -4     7
>> x^2
Error using  ^
Inputs must be a scalar and a square matrix.
To compute elementwise POWER, use POWER (.^) instead.
>> A^2
ans =
     7    10
    15    22
```

In order to do calculations component-wise on an array, use the "." before the operator. Thus ".*" is component-wise multiplication, ".^2" will square every component, etc.

```
>>
x.*y
ans =
     4    10    18
>> x.^2
ans =
     1     4     9
>> A = [1 2 3;4 5 6;-7 -8 -9]
A =
     1     2     3
     4     5     6
    -7    -8    -9
>> A^2
ans =
```

```
   -12   -12   -12
   -18   -15   -12
    24    18    12
>> A.^2
ans =
     1     4     9
    16    25    36
    49    64    81
```

Note that, for scalar multiplication, the "`.*`" is not necessary, but `3.*z` or `z.*3` will still work. For addition and subtraction, it will only work if the scalar is first.

```
>> 3.*z
ans =
    30    33
>> z.*3
ans =
    30    33
>> 3.+z
ans =
    13    14
>> z.+3
 z.+3
   |
Error: Unexpected MATLAB operator.
>> x.+y
 x.+y
   |
Error: Unexpected MATLAB operator.
```

Mathematical functions on vectors and matrices are automatically done component-wise.

```
>> abs(A)
ans =
     1     2     3
     4     5     6
     7     8     9
>> exp(A)
ans =
    2.7183    7.3891   20.0855
   54.5982  148.4132  403.4288
    0.0009    0.0003    0.0001
>> sind([0 30 45 60 90])
ans =
         0    0.5000    0.7071    0.8660    1.0000
```

Other useful functions

1. `length`
2. `size`
3. `numel`
4. `end`
5. `sort`
6. `max`
7. `min`
8. `sum`
9. `ones`
10. `zeros`
11. `eye`
12. `reshape`
13. and many others!

2.5. Random numbers

There are many uses for generating random numbers, vectors, or matrices. The command `rand` will return a uniformly distributed pseudorandom number between 0 and 1. The command `randn` is similar, except that it is normally distributed. To generate a random *integer* between 1 and k, use `randi(k)`. To permute the numbers from 1 to n, use `randperm(n)`.

```
>> x=rand
x =
    0.6348
>> y=randn
y =
    0.4598
>> z=randi(10)
z =
     6
>> p=randperm(5)
p =
     5     3     2     4     1
```

All of the above commands can be modified to generate random vectors and matrices, and other modifications can be made to `randi` and `randperm`. See Table 2.1 for a summary of the basic modifications. See the MATLAB documentation about fancier modifications, including information on the random number generator.

```
>> x=rand, y=randn, z=randi(10)
x =
```

```
    0.1492
y =
    0.7923
z =
     6
>> p=randperm(5)
p =
     5     3     1     2     4
>> x2=rand(2), y2=randn(2,3)
x2 =
    0.1903    0.2686
    0.0302    0.9827
y2 =
   -1.7616   -0.4471    0.1665
   -0.9166    0.1737   -1.0049
>> z2=randi(10,2), z3=randi(10,2,3), z4=randi([-10,10],1,4)
z2 =
     7     3
     8     9
z3 =
     9     1     2
     9     5     7
z4 =
    -9     1     6    -7
>> p2=randperm(10,3)
p2 =
    10     2     7
>> p3=randperm(10,3), z6=randi(10,1,3)
p3 =
     1     5     2
z6 =
     5     3     3
```

Note that we can modify `rand` to generate a random number between any two numbers and we can modify `randn` to have a different mean and standard deviation than the standard normal distribution. See examples below.

Example 2.5.1. Create 1000 random numbers between 5 and 8.

```
>> x=rand(1000,1)*(3) + 5;
>> histogram(x)
```

In Fig. 2.1 the command `histogram` is used to quickly display the generated random numbers.

Table 2.1 Useful random number generators.

`rand`	Uniformly distributed number between 0 and 1
`rand(n)`	$n \times n$ matrix of random numbers
`rand(m,n)`	$m \times n$ matrix of random numbers
`randn`	Normally distributed number ($\mathcal{N}(0, 1)$)
`randn(n)`	$n \times n$ matrix of normally distributed numbers
`randn(m,n)`	$m \times n$ matrix of normally distributed numbers
`randi(k)`	Uniformly distributed integer between 1 and k
`randi([a,b])`	Integer between a and b
`randi(k,n)`	$n \times n$ matrix of integers between 1 and k
`randi([a,b],n)`	$n \times n$ matrix of integers between a and b
`randi(k,m,n)`	$m \times n$ matrix of integers between 1 and k
`randi([a,b],m,n)`	$m \times n$ matrix of integers between a and b
`randperm(n)`	Random permutation of integers from 1 to n
`randperm(n,k)`	Random permutation of k *unique* integers from 1 to n

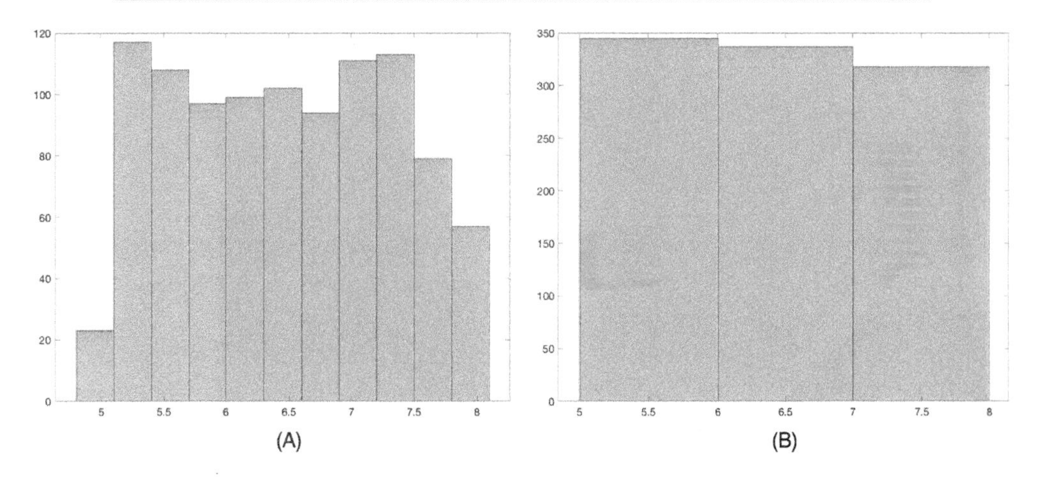

Figure 2.1 Visualizing `rand` modification. (A) `histogram(x)`, (B) `histogram(x, [5,6,7,8])`.

Example 2.5.2. Create 10,000 random numbers normally distributed with mean of 100 and standard deviation of 25. Note that our sampling of 10,000 data points from $\mathcal{N}(100, 25)$ would not give a mean and standard deviation of exactly 100 and 25, respectively.

```
>> y=randn(10000,1)*25 + 100;
>> [mean(y), std(y)]
ans =
   99.7555   25.0802
>> histogram(y)
```

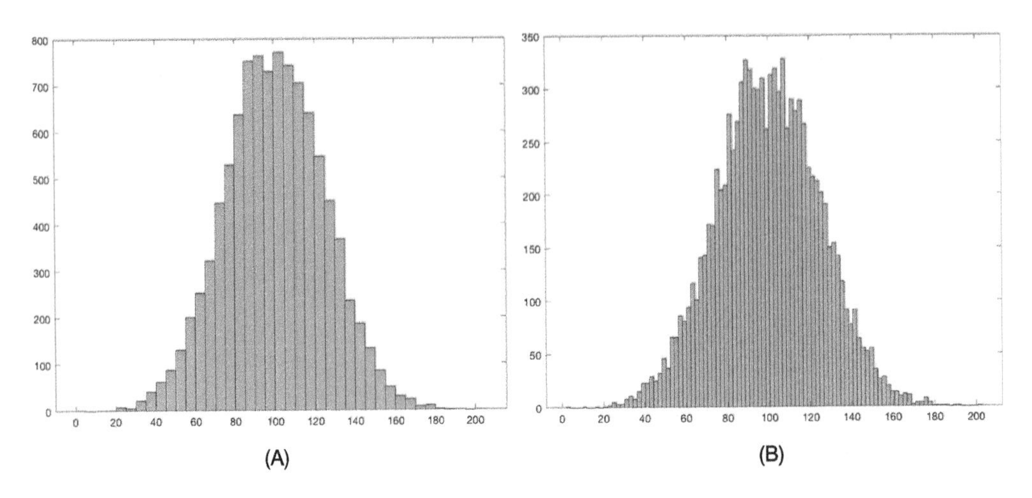

Figure 2.2 Visualizing `randn` modification. (A) `histogram(y)`, (B) `histogram(y,100)`.

In Fig. 2.2 the command `histogram` is used to display these normally distributed random numbers.

2.6. Exercises

NOTE: additional exercises that use vectors and/or matrices appear in Appendix C.

1. Do not use MATLAB help or documentation to answer these questions; only experiment with the commands.
   - **(a)** Using the `rand` command, create a random matrix `A1` with 4 rows and 6 columns.
   - **(b)** Create a matrix `B1` that is the transpose of `A1` (there are multiple ways to do this with one command; choose one).
   - **(c)** Using the `randi` command, create a random ROW VECTOR named `u1` with eight elements of integers from 1 to 100.
   - **(d)** Create a random COLUMN VECTOR named `v1` with nine elements of integers from −10 to 10.
   - **(e)** By using the MATLAB functions `length`, `size`, and `numel` on these matrices and vectors, figure out the differences between these functions, and come to your own conclusions as to when these functions would be useful. Write your conclusions nicely on a separate sheet of paper. Write complete sentences so each part that you are answering is clear. You will be graded on following directions and not on the accuracy of the descriptions of the functions, but whether you experimented sufficiently and your conclusions reflect your experimentation.

2. Do not use MATLAB help or documentation to answer these questions; only experiment with the commands.
   - **(a)** Create a vector `v2` with the numbers 2.1, 2.5, 2.6, −2.1, −2.5, and −2.6.
   - **(b)** By using the MATLAB functions `ceil`, `floor`, `fix`, and `round` on `v2` explain in your own words on paper what the functions do, making clear what the differences are between them. Just as above, you will be graded not on the accuracy of the descriptions of the functions but whether your experimented sufficiently and your conclusions reflect your experimentation.
3. Use the colon operator to generate the following vectors (do not suppress the output):
   - **(a)** `u3 = -5 -4 -3 -2 -1 0 1 2 3 4 5`
   - **(b)** `v3 = 0 0.2500 0.5000 0.7500 1.0000`
   - **(c)** `w3 = 12 9 6 3 0`
4. Create these vectors and matrices as efficiently as possible.
   - **(a)** Create a vector x that has nine equally spaced values between −2 and 2.
   - **(b)** Create a matrix `A4` that has the values of x as the first column, the values of $3x$ as the second column, and the values of $4x - 7$ as the third column.
   - **(c)** Create a matrix `B4` whose first row has the values πx and the second row has values that are cosine of the corresponding values of πx.
5. Create these vectors and matrices as efficiently as possible.
   - **(a)** Create the 4×4 identity matrix called `M`.
   - **(b)** Create a 10×10 matrix `T` filled with 1s except a 0 in the third row, fourth column position.
   - **(c)** Create a 8×9 matrix `F` filled with 0s except a 1 in the fifth row, seventh column position.
   - **(d)** Create a vector called `lengthu` that stores the length of the vector `u3` created above.
   - **(e)** Create variables `r` and `c` that store the number of rows and columns of the matrix `F` above, respectively.
6. Each row in the matrix $M6$ follows a different mathematical pattern. Using these patterns, enter the following matrix into MATLAB. Thus you should be able to do this without entering each value by hand. Using $M6$, do parts (a)–(d) as efficiently as possible:

$$M6 = \begin{bmatrix} 1 & 2 & 3 & 4 \\ 9 & 8 & 7 & 6 \\ 1 & 4 & 9 & 16 \\ -1 & 1 & -1 & 1 \end{bmatrix}.$$

   - **(a)** Create a vector `A6` consisting of the elements of the third column of $M6$ and a vector `W6` consisting of the elements of the second row of $M6$.

(b) Create a 4×3 matrix `B6` consisting of all elements of the second through fourth columns of $M6$.

(c) Create a 3×4 matrix `C6` consisting of all elements in the first through third rows of $M6$.

(d) Create a 2×3 matrix `D6` consisting of all elements of the second and third rows and last three columns of $M6$.

7. Use the matrix $M6$ above to find the following using MATLAB functions and commands as efficiently as possible. Think what you would need to use if $M6$ had 1000 rows and columns.

(a) Find the minimum values in each column of $M6$ and store the answers in the variable called `mincols`.

(b) Find the maximum values in each row of $M6$ and the location of these values. State your answers on your sheet of paper nicely. For example, "Maximum of row 1 = ___ and is in the i,j entry of M6."

(c) Sort each column of $M6$ and store the result in the matrix `M6colsort` and sort each row of $M6$ and store the result in the matrix `M6rowsort`. NOTE: there is a way to have MATLAB do this work for you without having to look at the matrix $M6$ yourself!

(d) Total the values in each column of $M6$ and store the result in `M6coladd` and total the values in each row of $M6$ and store the result in `M6rowadd`.

8. Use the matrices C and D below to find the following in MATLAB:

$$C = \begin{bmatrix} 11 & 5 \\ -9 & 4 \end{bmatrix}, \qquad D = \begin{bmatrix} -7 & -8 \\ -9 & -4 \end{bmatrix}.$$

(a) $E = CD$, $F = DC$, and $G =$ each element of C multiplied with its corresponding element of D.

(b) $H = C + D$, $J = C^2$, and $K =$ each element of C squared.

9. Try to do these as efficiently as possible.

(a) Using `rand`, generate a random real number between -2π and 2π and store it in the variable `prob9a` (do not suppress output).

(b) Using `randi`, simulate a roll of six dice for the game of Farkle by creating a vector of six random integers between 1 and 6 (so repeats are allowed) and store it in the variable `farkle` (do not suppress output).

(c) Generate a random order for 20 presentations (thus repeats are not allowed, unlike above) by using the `randperm` function if the presenters are numbered 1 through 20 (do not suppress output).

(d) Generate a matrix called `PowerballTix` that will simulate ten Powerball tickets/drawings (each row will be one drawing). The first five entries of each row are the "white balls" and the last entry of each row is the "powerball" for the drawing. For each row/drawing, the first five columns should be random integers from 1 to 69 **with no repeats** and not necessarily in order and the last column should be a random integer from 1 to 26 [3] (do not suppress output).

(e) A CDC study [7, p. 14] in the years 2011 to 2014 showed American females aged 20–29 had a mean height of 64.1 inches with a standard error of 0.12 inches. Using these numbers as the mean and standard deviation of a normal distribution, generate one random height and store it in a variable called `randomheight` (do not suppress output).

10. We will use the command `histogram` along with the random generator functions used above to visualize some of the data.

(a) Create a vector called `tenrolls` that will store ten rolls of a standard die (suppress the output!) and then use the `histogram` command to display the outcome of those ten rolls. Use six "nbins" for your histogram.

(b) Do the same as in part (a) but have the vector called `manyrolls` for 100,000 rolls of a standard die.

(c) The GRE guide [5, p. 18] reports that between July 1, 2015 and June 30, 2018 there were 1,695,463 test takers that took the Quantitative Reasoning part of the GRE with a mean of 153.07 and standard deviation 9.24. Simulate this normal distribution by generating a vector called `GRE` that has the same number of randomly generated scores as reported test takers (suppress this output!). Then use the `histogram` command to display your vector, using 100 "nbins" for your histogram.

11. Use the matrix $M6$ in #6 above to answer the following problems. Show the commands and/or experimentation that helped you reach your conclusions.

(a) What is the difference between `sum(M6)`, `sum(M6,2)`, `sum(M6,1)`, `sum(M6')`, and `sum(M6, 'all')`? Can you come up with an equivalent but different command for `sum(M6,2)`?

(b) What is the difference between the commands `sort(M6)`, `sort(M6,1)`, and `sort(M6,2)`?

(c) Can you use the same idea for `min` and `max`? In other words, do you get similar results for the commands `min(M6)`, `min(M6,2)`, `min(M6,'all')`, etc. as you do with the `sum` or `sort` command? If not, what is going on with these commands and how can you get the same results as in `sum(M6,2)`?

12. **(a)** Generate a vector named `theta1` of values from 0 to 2π with increments of $\pi/6$ using `linspace`. *(Hint: how many values should you have?)*

(b) Generate a vector named `theta2` of values from 0 to 2π with increments of $\pi/6$ WITHOUT using `linspace`.

(c) Calculate the sine of these angles using either `theta1` or `theta2`.

(d) Generate a table of Cartesian coordinates corresponding to these angles on the unit circle by building a matrix called `Cart`. The first column should be either `theta1` or `theta2`, the second column should be the x-coordinate, and the third column should be the y-coordinate of the points on the unit circle.

(e) Add a fourth column that contains the angle measurements in degrees and call this new matrix `Cart2`.

(f) Add rows (in the appropriate spots!) corresponding to $\pi/4$, $3\pi/4$, $5\pi/4$ and $7\pi/4$ and call this new matrix `Cart3` (this may take several steps).

CHAPTER 3

Plotting in MATLAB®

3.1. Basic 2D plots

There are many ways to plot data, curves, and/or functions. We will not cover all plotting commands, but hopefully after this chapter you will be able to use other plotting commands (such as `pie`, etc.) and be able to customize your plots easier.

We will start with plotting functions. If you are familiar with the command `ezplot`, this command is outdated and should be replaced by `fplot`, so we will not discuss `ezplot`. Both `ezplot` and `fplot` are useful but do not have as much versatility as the `plot` command. With `fplot` you can control the domain, colors, markers, etc. as we will discuss in this chapter but MATLAB® decides which data points to use within your domain to create the plot. For example, notice what the commands `fplot(@(x)x.^3)`, `fplot(@(x)sin(x))`, and `fplot(@(x)sin(x), [-2,2])` create.

With the command `plot` we can control which points, and thus how many points are used to create the plot(s). We can also plot data points. Most of the examples in this chapter will use `plot`. Much of the syntax for modifying colors, creating multiple plots, etc. are the same with `fplot` and `plot`.

In order to plot a mathematical function using `plot`, follow these steps.

Step 1: define the domain as a vector. This is where the `linspace` command can come in handy, and when you will want to SUPPRESS THE OUTPUT.

```
x = linspace(-5,5);
```

Step 2: calculate the y-values for each of the x-values in your domain USING COMPONENT-WISE CALCULATIONS. Note that `fplot` needs component-wise calculations as well.

```
y = x.^3;
```

Step 3: Plot the inputs and outputs using the `plot` command. What the `plot(x,y)` command does is plot each point (x_k, y_k) in the vectors `x` and `y` (or whatever you have called these variables). By default, it connects these points in the order of the vectors with a blue line. If there are enough points in the vectors, you get a smooth curve as expected in this case.

We will have to add labels and titles if we want them (shown later). One can also just as easily define the vectors within the `plot` command. For example, the following code would produce the same figure as in Fig. 3.1:

```
x=linspace(-5,5);
plot(x,x.^3)
```

Programming Mathematics Using MATLAB®
https://doi.org/10.1016/B978-0-12-817799-0.00008-9

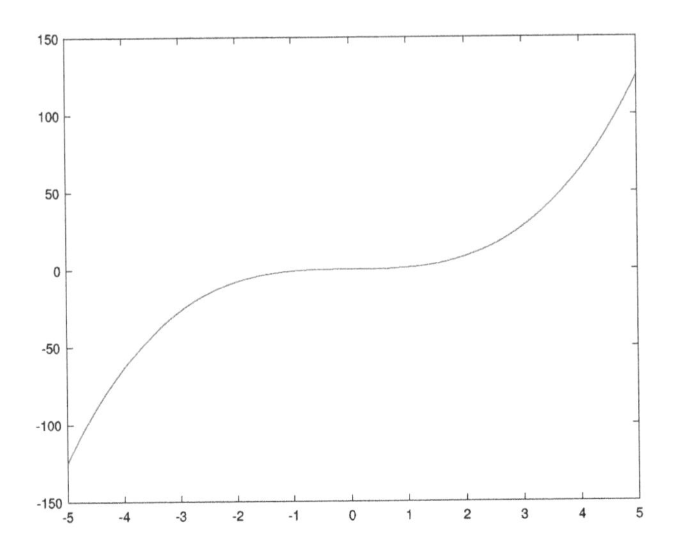

Figure 3.1 `plot(x,y)`.

3.2. Bad domain examples

Remember, MATLAB will plot points as they are defined in the vectors. Thus if the vectors do not have enough points, or they are in a certain order, the pictures may not come out as planned.

What we could do is make sure `x` has more elements, either by using `linspace` (oftentimes preferable), or by making the increment used in colon operator be small. Be careful: too many elements in the vector (from having the `n` in `linspace(a,b,n)` too large, or having too small of an increment if using colon operator) can cause MATLAB to slow down, especially when defining domains used in 3D plots.

The following have different domain definitions, followed by the commands `y=sin (x); plot(x,y)`. Notice the difference in the domain definitions, and thus the figures shown in Fig. 3.2.

Another issue is having the points in the wrong order. If we use the commands below to create a square, we see in Fig. 3.3(A) that the vertices are listed in the wrong order.

```
x=[0 0 1 1]; y=[0 1 0 1];
plot(x,y)
axis([-0.2 1.2 -0.2 1.2])
```

Notice also the use of the `axis` command so that the plot can be seen better. We must make sure the vertices are in the proper order so that when they are connected, we get the desired effect.

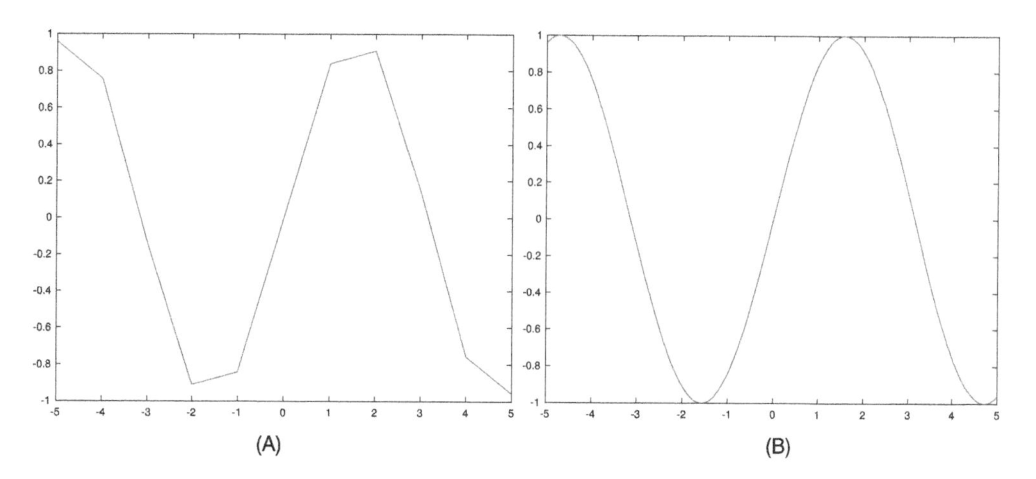

Figure 3.2 Domain examples. (A) Bad domain: `x = -5:5`, (B) Better domain: `x = -5:0.01:5`.

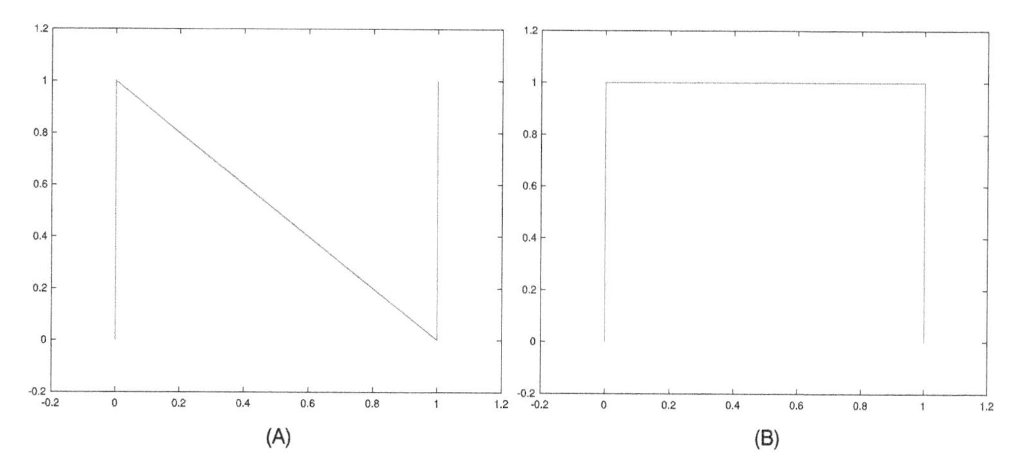

Figure 3.3 Reordering points. (A) `x=[0 0 1 1]; y=[0 1 0 1]`, (B) `x=[0 0 1 1]; y=[0 1 1 0]`.

We can make this even better by remembering to include the first point as the last point so all vertices are connected. Also, for Fig. 3.4(B), we added the command `axis equal` after the `axis([-0.2 1.2 -0.2 1.2])` command so that what is shown does indeed look like a square as desired.

3.3. Axis settings

There are many settings and properties of the axis figure. Not all will be covered here. The reader is directed to the MATLAB documentation for a more thorough treatment.

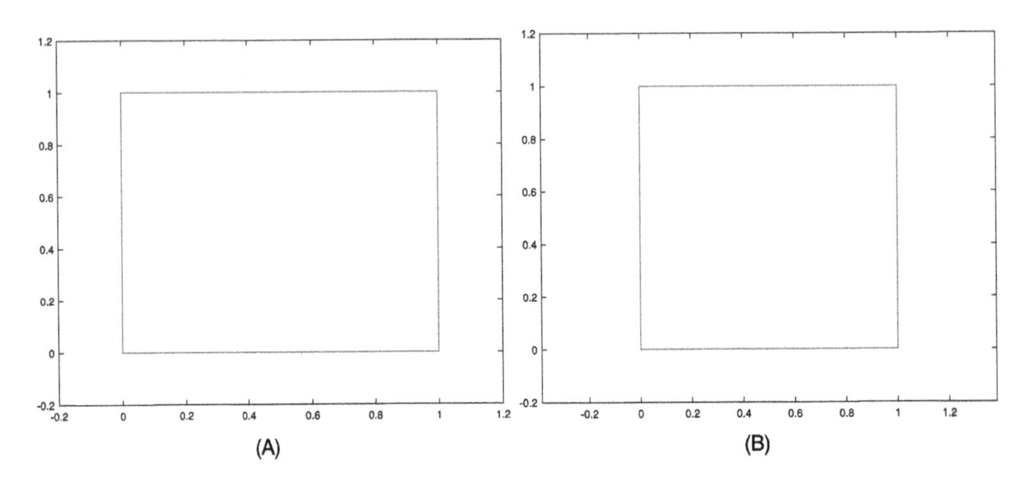

Figure 3.4 Better squares. (A) `x=[0 0 1 1 0]; y=[0 1 0 1 0]`, (B) Using `axis equal`.

As seen in Figs. 3.3 and 3.4, you can control the length of the axes shown and/or aspect ratios. You can even set `axis off`. You can change the limits of the x-values or y-values shown in the figure by using `xlim` and/or `ylim` commands. Fig. 3.12 uses the `xlim` command. You can also use the `axis` command. The `axis` command is in the form `[xmin xmax ymin ymax]`, with the addition of `zmin zmax` elements for a three-dimensional figure (discussed in Chapter 4). Thus `axis([-2 2 -5 5])` will set the axes to be $[-2, 2] \times [-5, 5]$. This command is equivalent to `xlim([-2,2]), ylim([-5,5])`.

To change the aspect ratio, especially when plotting circles, squares, etc. you may want to use `axis equal` or `axis square`. Another useful one is `axis tight`. To reset to the default, use `axis normal`. In Fig. 3.5 are different examples showing the differences between some of the `axis` settings to graph $y = \sin(x)$ for $x \in [-2\pi, 2\pi]$. A word of advice: if you are expecting lines and/or planes to look orthogonal (perpendicular) and they do not, it could be the aspect ratio of the figure. Learn from me, first try `axis equal` before spending too much time double checking and triple checking your math! This is demonstrated in Section 4.4.1.

To see the importance of setting the aspect ratio using `axis`, consider the following MATLAB code that generates Fig. 3.6(A):

```
t=linspace(0,2*pi);
x=cos(t); y=sin(t);
plot(x,y)
```

If you look closely, this is a parameterization of the unit circle, and yet the figure looks more like an ellipse. If we use the command `axis([-1.5 1.5 -1.5 1.5])` to

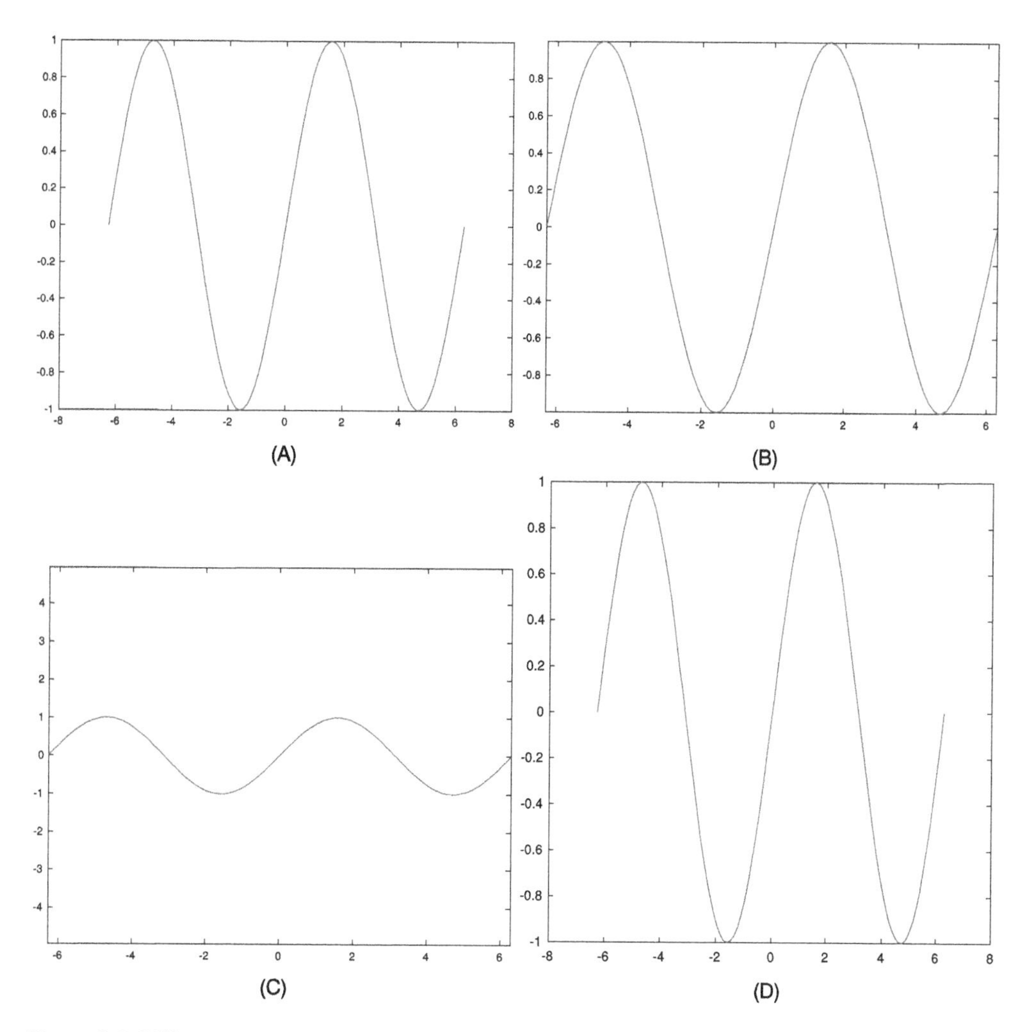

Figure 3.5 Different `axis` settings with sine. (A) Default axis, (B) `axis tight`, (C) `axis equal` and (D) `axis square`.

establish that `[xmin xmax ymin ymax]` should have values `[-1.5 1.5 -1.5 1.5]`, it still looks like an ellipse (Fig. 3.6(B)).

Setting the axis limits in Fig. 3.6(B) still does not achieve the desired picture. Other `axis` properties then need to be set, such as `axis square` and `axis equal` (Fig. 3.7).

A word of caution: when you are combining both setting the limits and the aspect ratios, **the order of these commands is important**! You may need to experiment with different orders to see the difference and which order gives you the desired effect. For example, in the $y = \sin(x)$ plot below, if you have the `axis equal` before or after

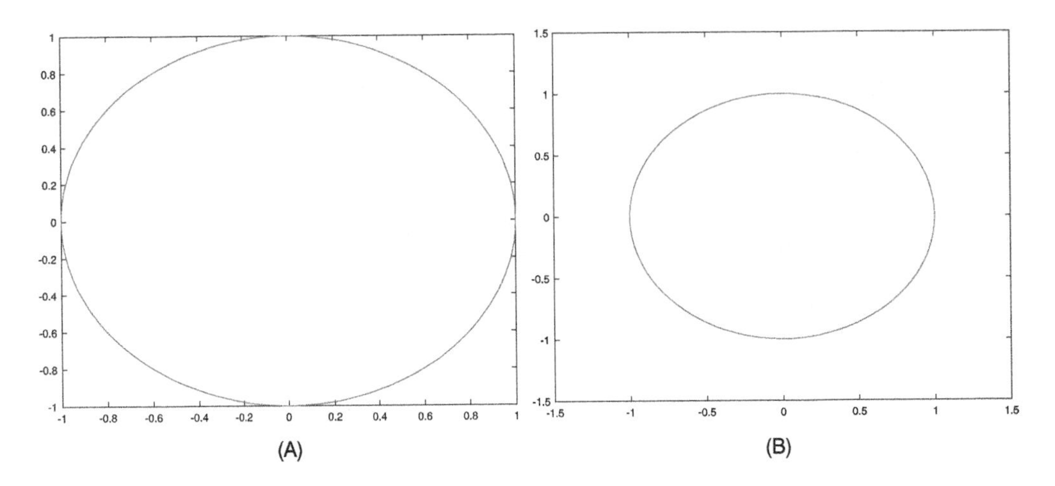

Figure 3.6 Unit circle. (A) Default axis and (B) `axis([-1.5 1.5 -1.5 1.5])`.

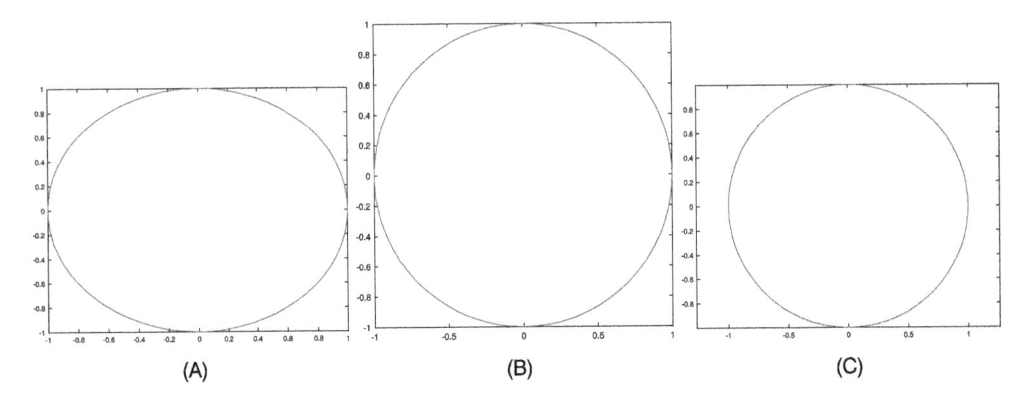

Figure 3.7 Axis Examples. (A) The default axis, (B) `axis square` and (C) `axis equal`.

the `ylim([-2,2])` commands, you get different results (see Fig. 3.8). Also, if you add plots after setting the axes, the axes are already "set" so the additional plot will not change the limits or aspect ratio.

```
x=linspace(-10,10);
y=sin(x);
plot(x,y)
axis equal
axis([-5, 5, -2,2])
```

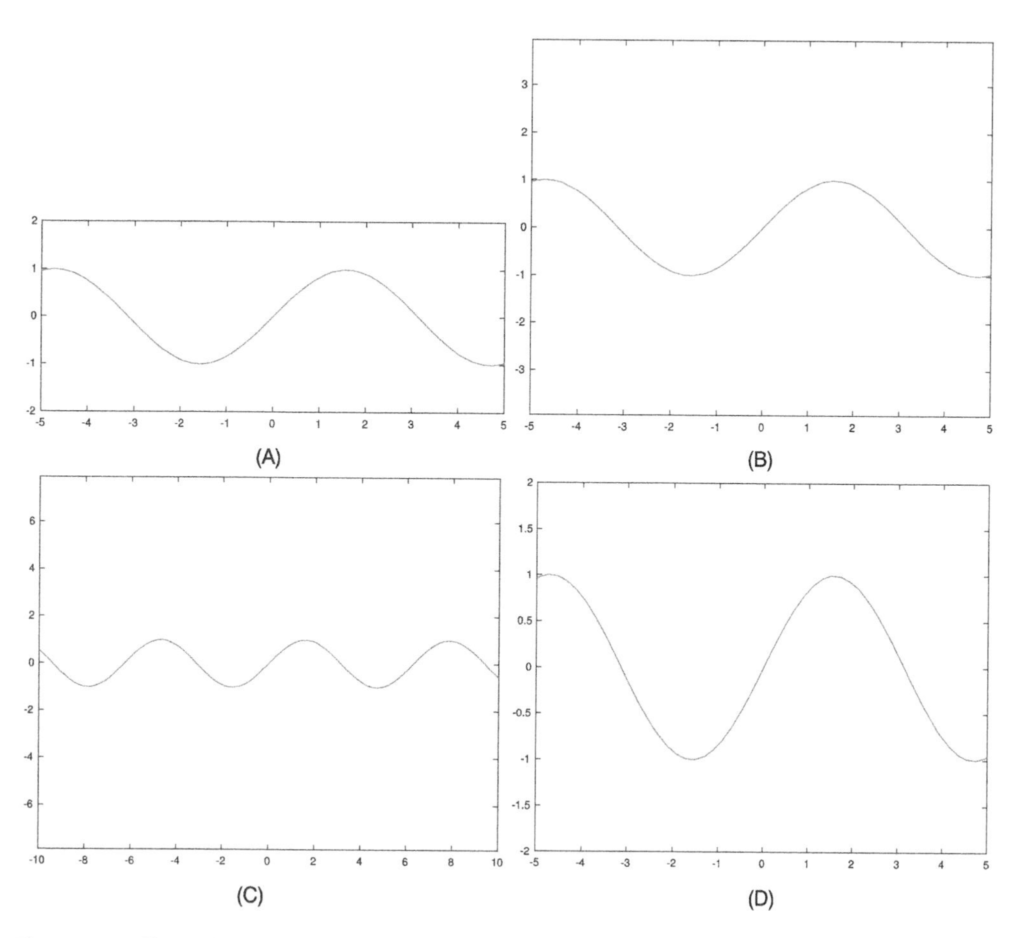

Figure 3.8 Effects of order of commands. (A) `axis equal` first, (B) `axis equal` second, (C) Only `axis equal` and (D) Only `axis([-5, 5, -2,2])`.

Experiment with plotting the circle, setting the axes from −1.5 to 1.5 and `axis equal`. Have both commands, in different orders, and just one of those commands to see the differences in the plots created. Experimenting you may notice the following:

```
axis([-1.5 1.5 -1.5 1.5]), axis square
%% is equivalent to
axis square, axis([-1.5 1.5 -1.5 1.5])
%% is equivalent to
axis equal, axis([-1.5 1.5 -1.5 1.5])
%% but is different from
axis([-1.5 1.5 -1.5 1.5]), axis equal
```

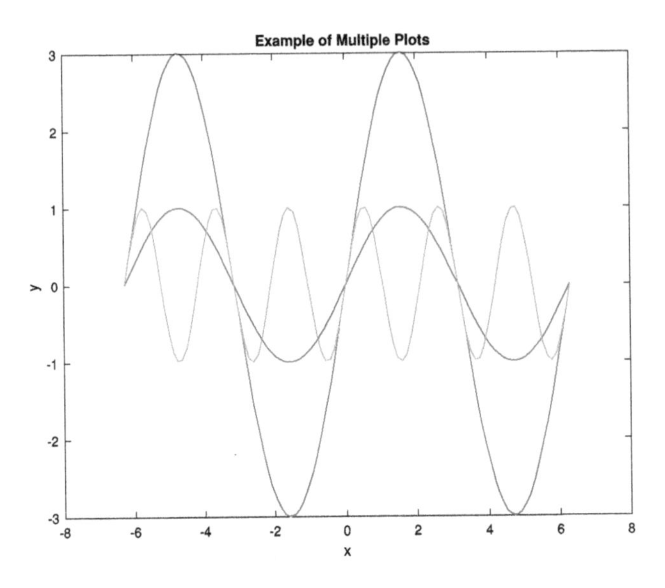

Figure 3.9 Multiple plots with one `plot` command.

3.4. Multiple plots

There are several ways you can put multiple plots within one figure. Keep in mind that whenever MATLAB sees a new plot command, whether it be `plot`, `scatter`, or `mesh`, etc., it replaces any previous plot in the active figure window with the new command unless we tell it otherwise.

The most basic way to have multiple plots is to put all of them within one command, as in Fig. 3.9. The drawback is that this can be more cumbersome when modifying how the curves/points look.

```
x = linspace(-2*pi,2*pi);
y = sin(x);
y2 = 3*sin(x);
y3 = sin(3*x);
plot(x,y,x,y2,x,y3)
xlabel('x'),ylabel('y')
title('Example 1 of Multiple Plots')
```

Notice that we also put in axes labels and a title in the above plot.

You can easily create a legend, making sure the text within the legends is in the same order as the order within the `plot` command. By comparing Figs. 3.9 and 3.10, notice the effect of the `axis tight` command. In Fig. 3.10(B), we have changed the location of the legend by using the command

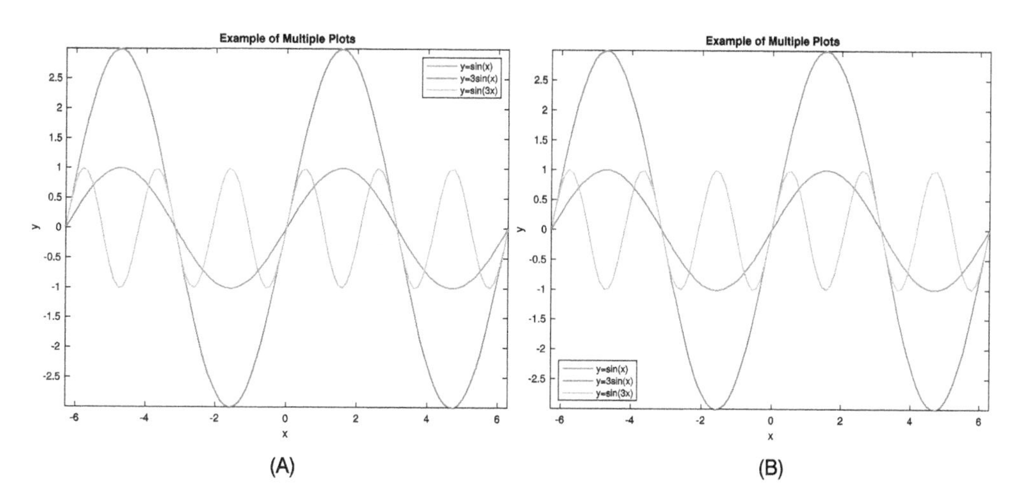

Figure 3.10 Using `legend` and `axis tight`. (A) Default `legend` location, (B) Changing location of legend.

```
legend('y=sin(x)', 'y=3sin(x)', 'y=sin(3x)', 'Location','SouthWest').
```

```
x = linspace(-2*pi,2*pi);
y = sin(x);
y2 = 3*sin(x);
y3 = sin(3*x);
plot(x,y,x,y2,x,y3)
xlabel('x'),ylabel('y')
title('Example of Multiple Plots')
legend('y=sin(x)', 'y=3sin(x)', 'y=sin(3x)')
axis tight
```

Another way to have multiple plots is to use the `hold on` and `hold off` commands. The `hold on` command tells MATLAB to hold the active figure window open to add to it with any subsequent plotting commands until the `hold off` command appears. Thus you can mix various plot commands in one figure window (when appropriate). For example, you can use the `plot` command and the `scatter` command on one figure, as in Fig. 3.11(A).

```
x = randi(10,1,50); % 1x50 vector of random integers from 1-10
y = randi(10,1,50); % create another similar vector
scatter(x,y)  % scatter plot of the points (x,y)
hold on
    plot(x,x)       % plot the line y=x
hold off
```

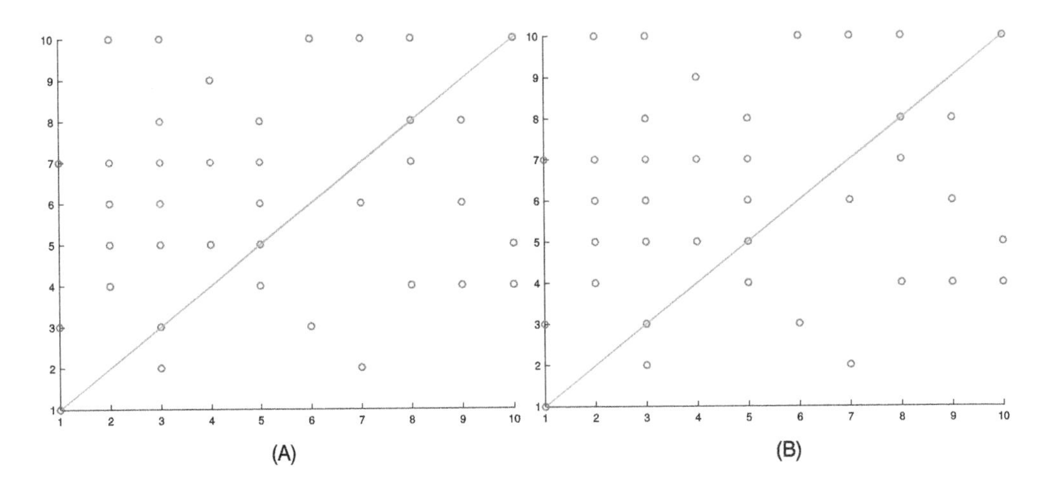

Figure 3.11 Using `hold on` and `hold off` command. (A) Using `scatter`, (B) Using `plot`.

Note that you can just as easily use the command `plot` instead of `scatter`, although for the `plot` command you must specify to just plot the points with a marker that you designate, rather than connect the points (see Fig. 3.11(B)).

```
plot(x,y, 'o')          % plot the points (x,y) using marker 'o'
hold on
    plot(x,x)           % plot the line y=x
hold off
```

When using `hold on` and `hold off` for multiple plots, the lines/markers will cycle through the default colors (new from MATLAB version R2014b on). The `grid on` command can also be useful.

```
x = linspace(-2*pi,2*pi);
y= sin(x);
y2=3*sin(x);
y3 = sin(3*x);
plot(x,y)
hold on
        plot(x,y2)
        plot(x,y3)
xlabel('x'),ylabel('y')
title('Example of Multiple Plots')
xlim([-4,4])
grid on
legend('y = sin x', 'y = 3sin x', 'y = sin(3x)', 'Location', 'NorthWestOutside')
hold off                                  %% DON'T FORGET TO HOLD OFF!!
```

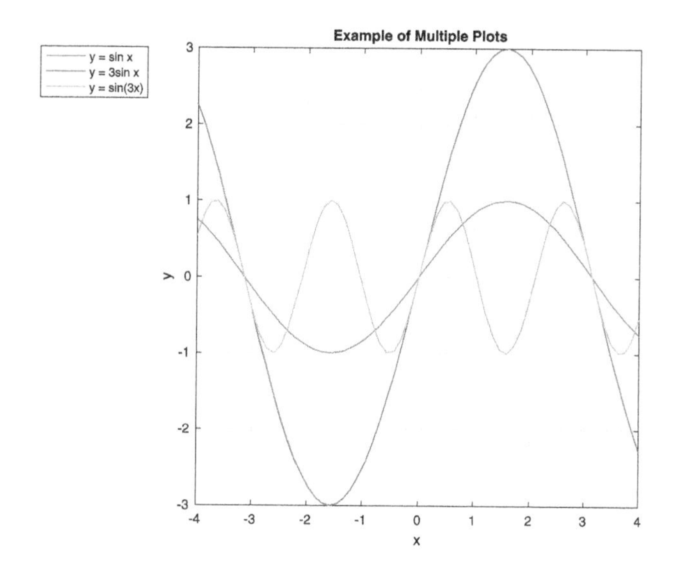

Figure 3.12 Using `hold on` and `hold off` command.

3.5. Color, line, and marker modifications

You can modify the solid line to dotted, dashed, etc., create markers at certain point(s), and/or adjust the size of the lines and/or markers as in Fig. 3.13.

```
plot(x,y,'g','LineWidth',2)            % line green, thicker
hold on
plot(x,y2,'r--', 'LineWidth',2)        % red, dashed, thick line
plot(x,y3,'b-')                                  % blue solid line (default)
plot(x,cos(x), 'k:o', 'LineWidth', 2) % black, dotted, thick
x2 = -2*pi:pi/4:2*pi;                        % creating points
y4 = sin(3*x2);
plot(x2,y4,'b*')                                 % blue * marker (no line)
plot(0,0,'kx', 'LineWidth',2,'MarkerSize',10) % big black x
xlabel('x'),ylabel('y')
title('Example of Multiple Plots')
legend('y = sin x','y = 3sin x','y = sin(3x)','y = cos(x)')
axis([-pi, pi, -5,5])
hold off %% DON'T FORGET TO HOLD OFF!!
```

Note that the **default first color is a different blue (`[0, 0.4470, 0.7410]`) from `'b'= [0, 0, 1]`, which is new from MATLAB version R2014b on. The default line style is solid, the default LineWidth is 0.5 and the default MarkerSize is 6**. Table 3.1 lists the possible basic colors, line styles, and markers. One, two, or three of these specifications can be defined in any order.

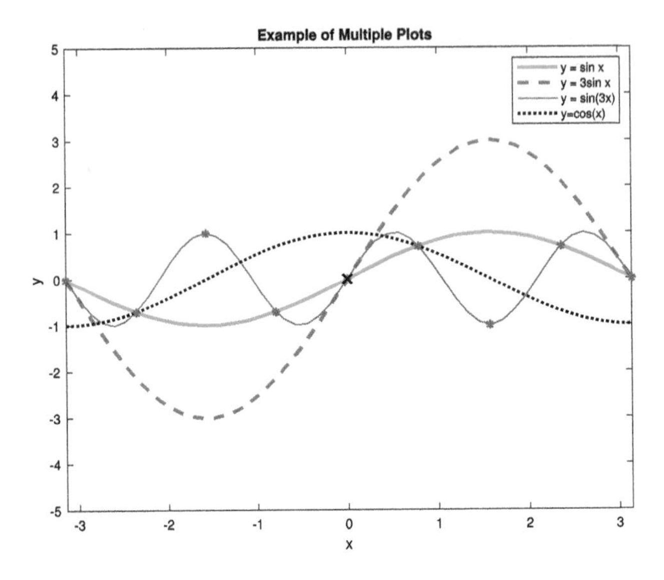

Figure 3.13 Adding markers and modifying lines/markers.

Table 3.1 List of basic colors, line styles, and markers.

Colors	'b'	blue	Markers	'+'	plus symbol
	'c'	cyan		'o'	open circle
	'g'	green		'*'	asterisk
	'k'	black		'x'	x
	'm'	magenta		's'	square
	'r'	red		'd'	diamond
	'w'	white		'p'	pentagram
	'y'	yellow		'h'	filled hexagram
Line Styles	'-'	solid (default)		'<'	left-pointing triangle
	'--'	dashed		'>'	right-pointing triangle
	':'	dotted		'^'	upward triangle
	'-.'	dash-dot		'v'	downward triangle

Note that you can use the "short name" shown above, the long name, or the RGB Value of these basic eight colors as shown in Table 3.2. Also the dashes, dash-dots, etc. may not appear nicely if you have too many points in the vectors plotted.

Consider Fig. 3.11. Because the default colors are cycled through (which is new starting in MATLAB version R2014b), the line and data points are different colors but we may want them to be the same color. One way is to specify the colors.

```
x = randi(10,1,50);
y = randi(10,1,50);
```

```
scatter(x,y, 'b')
hold on
    plot(x,x, 'b')
hold off
```

Another way is to reset to the first default color as is done in the following code:

```
scatter(x,y)
hold on
ax = gca;
ax.ColorOrderIndex = 1; % resets back to first color for next plot
plot(x,x)
hold off
```

The `lines` command is very useful to capture the colors of the current color map. Since there are 7 different colors, `lines(7)` will return a matrix where each row is the RGB code for the colors. Thus you can store those and assign them. The code below would return a similar plot as in the code above.

```
c=lines(7); % stores 7 default colors
c1=c(1,:); % assigns first color
scatter(x,y)
hold on
plot(x,x, 'Color', c1)
hold off
```

Another way is to specify different colors other than the basic eight colors discussed above is to specify colors by their RGB triple (for some examples, see http://www.rapidtables.com/web/color/RGB_Color.html). While many RGB colors are triples with numbers from 0 to 255 where (000) is black and (255, 255, 255) is white, MATLAB expects each number in the triple to be a number between 0 and 1. Thus you can take the RGB triple (255, 69, 0) for the color "orange red" and define it in MATLAB as `OrangeRed = 1/255*[255,69,0];`. Likewise, you can define the color

Table 3.2 Basic eight colors.

Short name	Long name	RGB value
'b'	blue	[0 0 1]
'c'	cyan	[0 1 1]
'g'	green	[0 1 0]
'k'	black	[0 0 0]
'm'	magenta	[1 0 1]
'r'	red	[1 0 0]
'w'	white	[1 1 1]
'y'	yellow	[1 1 0]

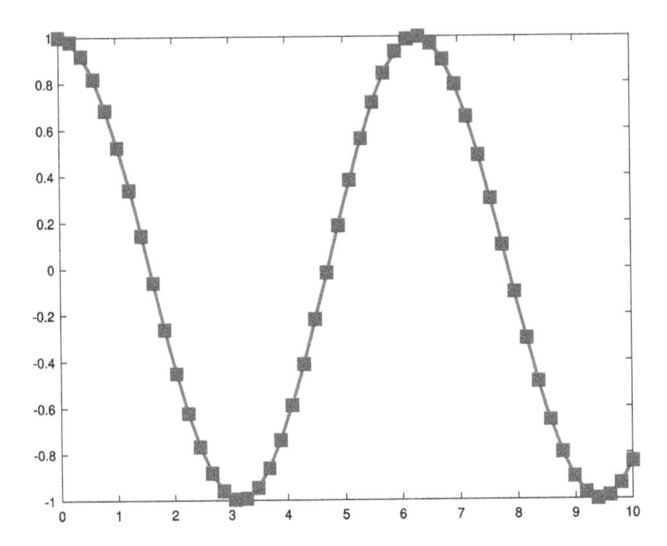

Figure 3.14 Using other colors.

`LoyolaGreen = 1/255*[0, 104, 87];` then one can use the defined color triple in your `plot` commands (see Fig. 3.14).

```
OrangeRed = 1/255*[255,69,0]; LoyolaGreen=1/255*[0, 104, 87];
x=linspace(0,10,50);
plot(x,cos(x),'-', 'color',OrangeRed,'LineWidth',2)
hold on
plot(x,cos(x),'s', 'MarkerFaceColor',LoyolaGreen,...
    'color',LoyolaGreen, 'LineWidth',2,'MarkerSize',10)
hold off
```

3.5.1 Clf/close all

The `clf` command stands for "clear figure". This will clear most figure settings such as `axis`, etc. that may have been applied to previous figure. If more than one figure window is open, it applies to the current or active figure window. It is also needed to clear the subplots. Note that some figure settings are not reset with the `clf` command. The command `close all` will close all figure windows. This will reset all figure commands, but can also slow things down so sometimes `clf` is preferred.

3.5.2 Subplots

The command `subplot(m,n,k)` is used to create a matrix of subplots within the same figure window. The number `m` is how many rows, `n` is how many columns, and `k` is which

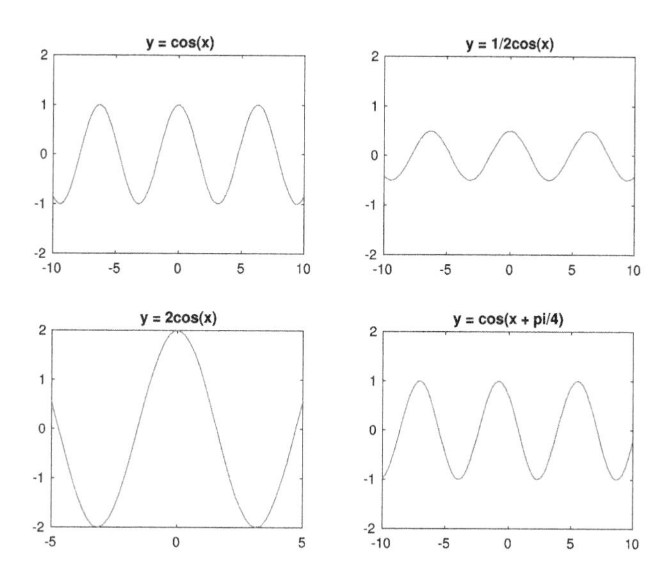

Figure 3.15 Subplot example.

subplot the following commands apply to. NOTE: you do not need to use commas to separate the m, n, and k. It is easiest to explain by viewing an example (see Fig. 3.15).

```
subplot(2,2,1)
    x = linspace(-10,10);
    y = cos(x);
    plot(x,y)
    title('y = cos(x)')
    ylim([-2,2])
subplot(2,2,2)
    y = 1/2*cos(x);
    plot(x,y)
    title('y = 1/2cos(x)')
    ylim([-2,2])
subplot(2,2,3)
    y = 2*cos(x);
    plot(x,y)
    title('y = 2cos(x)')
    axis([-5 5 -2, 2])
subplot(2,2,4)
    y = cos(x + pi/4);
    plot(x,y)
    title('y = cos(x + pi/4)')
    ylim([-2, 2])
```

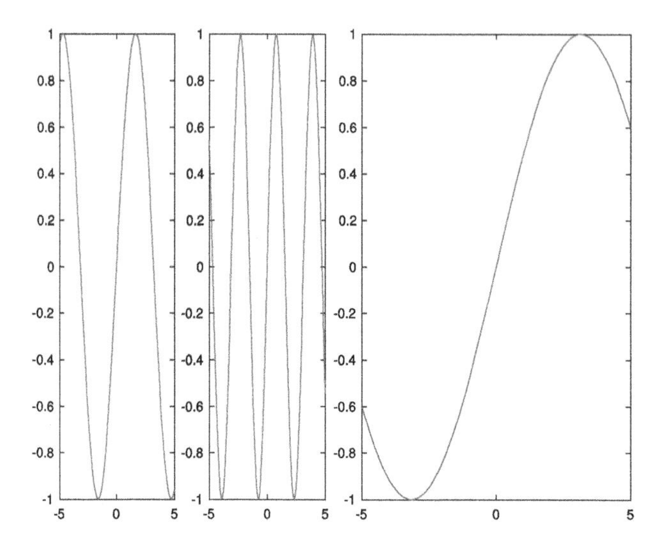

Figure 3.16 Subplot stretching across elements.

Notice that the numbering goes across the rows, first, then down. There is no way to make a "global title" without using the `text` command, and there it is not as nice because of having to specify the exact placement of the text.

One can have a subplot stretch across elements (see Fig. 3.16):

```
clf
subplot(141)
    x=linspace(-5,5);
    y=sin(x);
    plot(x,y)
subplot(142)
    y = sin(2*x);
    plot(x,y)
subplot(1,4,3:4)
    y=sin(1/2*x);
    plot(x,y)
```

Issue with subplots: once you have used `subplot` commands, any plotting commands will apply to the subplot currently set. Thus if I now have enter additional plotting commands, it will place it in the last subplot, in this case it will place it in the 3-4 spot from the above. Here is an example. If this code immediately follows the above code, the result is Fig. 3.17. Thus one should use the command `clf` or `close all` between subplots and subsequent plots to "reset" things.

```
plot(x,-abs(x))
```

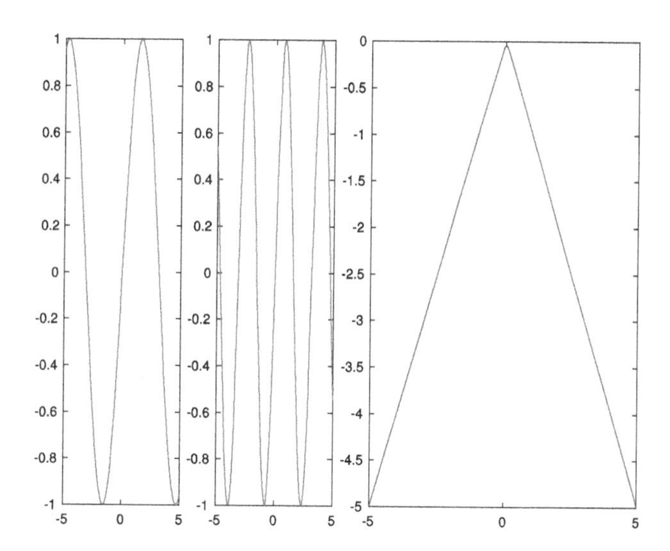

Figure 3.17 Not clearing figure after subplot.

One can use loops (discussed in Chapter 6) with subplots.

3.6. Other 2D plots

One nice thing about understanding the `plot` command is you can easily plot curves in which y is not necessarily a function of x. For example, we can easily plot the curve $x = y^2 + 2y - 3$, as shown in Fig. 3.18.

```
y = linspace(-5,5);
x=y.^2+2*y-3;
plot(x,y)
xlabel('x'),ylabel('y')
```

3.6.1 Parametric curves

You can also easily graph parametric equations such as

$$x = 3\cos t,$$
$$y = 2\sin t,$$

for $t \in [0, 2\pi]$.

```
t = linspace(0,2*pi);
x = 3*cos(t);
```

```
y = 2*sin(t);
plot(x,y)
xlabel('x'),ylabel('y')
title('Example of Parametric Equations')
axis equal
axis([-5 5 -5 5])
grid on
```

3.6.2 Polar curves

We can plot polar equations as parametric equations using the conversion equations. Notice in Fig. 3.19(A) that the curve appears somewhat jagged. This is because we are connecting only 100 points for the curve. We can create a smoother curve, by increasing the number of points in our `linspace` command. Fig. 3.19(B) was created by defining `t = linspace(0,2*pi,300);`.

```
t = linspace(0,2*pi);
r = 1 + sin(10*t);
x = r.*cos(t);
y = r.*sin(t);
plot(x,y)
xlabel('x'),ylabel('y')
```

Instead of using the conversion equations, one can have MATLAB do the conversion using the command `pol2cart`. You can have a polar axis with the command `polarplot`

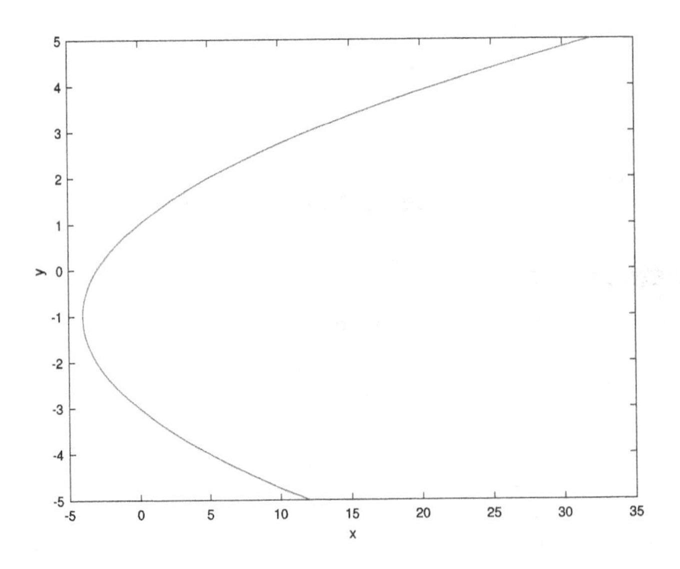

Figure 3.18 Other plots.

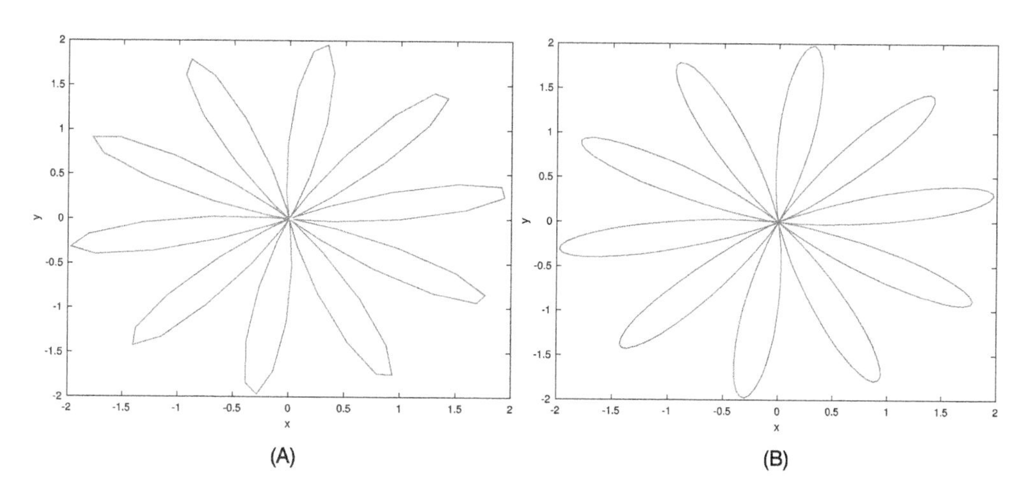

Figure 3.19 Polar curves. (A) Default `linspace`, (B) 300 elements.

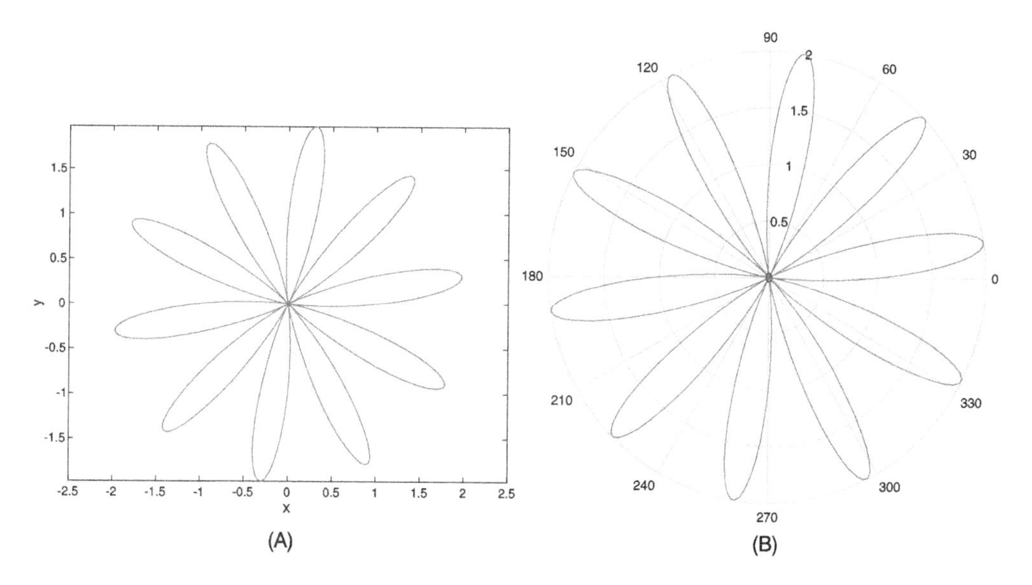

Figure 3.20 Polar curves. (A) With `pol2cart` (A) and (B) with `polarplot`.

(see Fig. 3.20). This command wants values for θ and r in polar coordinates. Note that the command `polar` is no longer advised to use.

```
%% using pol2cart
t = linspace(0,2*pi,300);
r = 1 + sin(10*t);
[x,y]=pol2cart(t,r);
plot(x,y)
```

```
xlabel('x'),ylabel('y')
axis equal
%% using polarplot
t = linspace(0,2*pi,300);
r = 1 + sin(10*t);
polarplot(t,r)
```

3.7. Exercises

1. The main span of the Golden Gate Bridge [9] can be roughly modeled with a catenary equation

$$y = 4200\cosh\left(\frac{x}{4346}\right) - 3954$$

for $|x| \leq 2100$. You can also model it with a quadratic over the same domain:

$$y = 0.00011338x^2 + 246.$$

   (a) Plot both of them on the same figure, making the catenary in red and the quadratic in black (default width). CAREFUL! How many elements should be in your domain?

   (b) Plot the catenary curve, making the line thicker than default and setting `'color'` [8,21] to be the RGB vector for an approximation to Golden Gate's International Orange:

$$\frac{1}{255}(155, 25, 0).$$

   Also use the command "`axis equal`".

2. The Gateway Arch in St. Louis was designed by Eero Saarinen and the central curve is modeled by a catenary

$$y = 693.8597 - 68.7672\cosh(0.0100333x),$$

with $|x| \leq 299.2239$ (x and y are both distances in feet) [20].

   (a) Plot the curve, making the line black.

   (b) Plot the curve again, making the line thicker than the default and setting `'color'` to be the RGB color for silver:

$$\frac{1}{255}(155, 25, 0).$$

   (c) What is the height of the arch at its center? State the exact value and use MATLAB to estimate the answer if needed (work on paper).

(d) At which point(s) is the height 100 m? State the exact value and use MATLAB to estimate the answer if needed (work on paper).

(e) What is the slope of the arch at the points in part (d)? Use MATLAB to estimate the answer if needed and show any work on paper.

(f) Create a second plot where the curve of the arch is shown, and the tangent lines at these points is shown in red. Mark the points with a black "x".

3. Consider the function $f(x) = \dfrac{2x^2 - 5x + 11}{x^2 - 7x - 8}$. We will work to make a decent graph (closer to one you may see in a textbook) in steps.

(a) Notice that the function has (a) vertical asymptote(s). What are they? State your answer as a full sentence, showing all work on paper. (Your answers should be equations for vertical lines!)

(b) This function also has horizontal asymptote(s). What are they? State your answer as a full sentence, showing all work on paper.

(c) Plot the function $y = f(x)$ using `fplot` without any other settings specified other than axes labels and a title.

(d) Plot the function $y = f(x)$ for $-20 \leq x \leq 20$ using `plot` without any other settings other than axes labels and a title. Notice this may not look like the way you may see the graph for this function in a textbook.

(e) Now use `plot` with $y = f(x)$ for $-20 \leq x \leq 20$ with some modifications. Set the range of the y-axis from -20 to 20. Notice again this may not look "textbook quality".

(f) Now use `plot` with $y = f(x)$ (same domain and range for the y as above) by dividing the domain into separate domains for each side of the vertical asymptotes. The graphs of each piece should all be the same color so that it looks like the same function.

(g) Do the same as in part (3f) but also graph the vertical asymptote(s) as red, dashed lines. Graph the horizontal asymptotes as black, dotted lines.

4. Consider the function $f(x) = \frac{1}{3}x^4 - 8x^2 + 4x + 1$.

(a) Using calculus, find the first and second derivatives of $f(x)$.

(b) Plot $y = f(x)$, $y = f'(x)$, $y = f''(x)$ on the same graph/figure for $x \in [-5, 5]$. Make sure you use a descriptive legend for each of the functions plotted.

(c) EXTRA CREDIT TWO POINTS: use `axis off` and create horizontal and vertical lines (in black) for the x and y axes. The axes should be of appropriate length for the graph. (There is a way to do this without knowing the height and width ahead of time!)

5. Consider the functions $f(x) = \cos(3x) - 1$ and $g(x) = 2x\sin(x)$.

(a) Using calculus, find the exact value of $\lim\limits_{x \to 0} \dfrac{f(x)}{g(x)}$ on paper, showing all work.

(b) To demonstrate the calculus, plot $y = f(x)/g(x)$, $y = f'(x)/g'(x)$, and $y = f''(x)/g''(x)$ near $x = 0$ on the same graph. You may have to adjust your axes to make it a nice looking graph and make sure you use a descriptive legend so one can tell which functions are which.

6. Consider the functions $f(x) = \sec(5x) - 1$ and $g(x) = 7x\sin(x)$.
   (a) Using calculus, find the exact value of $\lim_{x \to 0} \frac{f(x)}{g(x)}$ on paper, showing all work.
   (b) To demonstrate the calculus, plot $y = f(x)/g(x)$, $y = f'(x)/g'(x)$, and $y = f''(x)/g''(x)$ near $x = 0$ on the same graph. You may have to adjust your axes to make it a nice looking graph and make sure you use a descriptive legend so one can tell which functions are which.
7. Create a script that displays the following. The amount $A(t)$ of an initial investment P in an account paying an annual interest rate r at time t is given by

$$A(t) = P\left(1 + \frac{r}{n}\right)^{nt}$$

where n is the number of times the interest is compounded in a year and t is the number of years. If the interest is compounded continuously, the amount is given by

$$A(t) = Pe^{rt}.$$

Consider an investment of \$7500 put into a trust-fund account at an annual interest rate of 2.5% for 21 years. Show the difference in the value of the account when the interest is compounded annually, quarterly, and continuously by plotting $A(t)$ from $t = 15$ to $t = 21$ for these three situations on the same figure. Use a different line type (color and/or type of line), label the axes, create a meaningful legend and title for the plot.

8. For a calculus problem, you want to plot the region enclosed by the curves

$$x = y^2 - 3, \quad x = e^y + 1, \quad y = -1, \quad y = 1.$$

Create a script file to plot these four curves, making the first two blue and red, respectively, and the last two curves in black. Set the window to be $[-5, 5] \times [-5, 5]$.

9. If a projectile is thrown or fired with an initial velocity of v_0 meters per second at an angle α above the horizontal and air resistance is assumed to be negligible, then its position after t seconds is given by the parametric equations

$$x = (v_0 \cos\alpha)t, \qquad y = (v_0 \sin\alpha)t - \frac{1}{2}gt^2,$$

where g is the acceleration due to gravity (9.8 m/s^2).
   (a) According to Guinness World Records, the fastest lacrosse shot was recorded at 53.6 m/s in 2015 [12]. Use this speed as initial velocity v_0, and $\alpha = 8.5°$ to answer the following:

i. When will the lacrosse ball hit the ground?
ii. A lacrosse field is 100 m long. If a player is standing at one end of the field, does the ball make it to or past the other end of the field? How far from the player will it hit the ground?
iii. What is the maximum height reached by the ball?
iv. Calculate these showing all work and graph the position of the ball to check and demonstrate your answers. Label the axes appropriately.

(b) Using `subplot`, graph the path of the ball for the original value of α and three other values (with appropriate titles!) demonstrating how your answers to part (a) may change.

(c) Using `subplot`, graph the path of the ball for the original value of v_0 and three other values (with appropriate titles!) demonstrating how your answers to part (a) may change.

10. We will use the command `histogram(y,nbins)` along with the random generator functions to visualize data. Some languages only have one random number generator similar to `rand` in MATLAB. Common ways to simulate a roll of the die is to multiply these numbers by 6 and then modify them in some way to get integers between 1 and 6. Notice that you can then adjust this method to simulate any sided die that you would like. There is an easier way to simulate dice rolls in MATLAB using `randi`. We will create vectors that will store 10,000 rolls of a standard die (suppress the output!) using different methods and then visually compare the methods with histograms.

(a) Create vectors called `diceRolls1`, `diceRolls2`, `diceRolls3`, and `diceRolls4` that will store 10,000 rolls of a standard die (suppress the output!).

- The vector `diceRolls1` will use `rand` and `ceil` appropriately.
- The vector `diceRolls2` will use `rand` and `floor` appropriately.
- The vector `diceRolls3` will use `rand` and `fix` appropriately.
- The vector `diceRolls4` will use `randi` appropriately.

(b) Using `subplot`, create a 2×2 grid of plots comparing the four methods using the `histogram` command to display the outcome of those rolls. Use six "nbins" for your histogram. Have titles in each of the four subplots to specify which method was used.

(c) Based on your histograms, is there a method that seems better, or a method that creates an "unfair" die, or are they about the same? You may want to rerun the section of code multiple times within MATLAB to be more confident of your answer.

(d) Why did we not have a method that uses `rand` and `round`? Could you create a fair die that would use `rand` and `round`? Support your answer by simulating 10,000 rolls and creating a histogram.

Table 3.3 Orbit data.

	$a\,(AU)$	e
Mercury	0.3871	0.2056
Venus	0.7233	0.0068
Earth	1.0000	0.0167
Mars	1.5273	0.0930
Neptune	30.0699	0.0090
Pluto	39.4869	0.2489

11. Plot the following polar plots on one figure using `subplot`. Be sure that the titles name the curves, or give the equation if it is unnamed. Also be sure to choose the correct domain for each to make sure for each to make sure you produce the entire curve. Plot these using the `pol2cart` command.

- **(a)** **Freeth's nephroid** $r = 1 + 2\sin(\theta/2)$;
- **(b)** **Hippopede** $r = \sqrt{1 - 0.8\sin^2\theta}$;
- **(c)** **Butterfly curve** [6] $r = e^{\sin\theta} - 2\cos(4\theta) + \sin^5(\theta/12)$, $\theta \in [0, 50\pi]$;
- **(d)** $r = \sin^2(4\theta) + \cos(4\theta)$;
- **(e)** $r = 2 + 6\cos(4\theta)$;
- **(f)** $r = 2 + 3\cos(5\theta)$.

12. Using Kepler's first law of planetary motion, one can obtain a basic model of the orbits of planets and dwarf planets around the sun can be modeled using the polar equation [24, p. 727]

$$r = \frac{a(1 - e^2)}{1 + e\cos(\theta)}$$

where a is the semi-major axis and e is the eccentricity of the planet. The values of several semi-major axes (a) and eccentricities (e) for our solar system [23] are in Table 3.3.

- **(a)** Plot the orbits of Mercury, Venus, Earth and Mars in one figure using `polarplot` on the polar plane, making sure to have a legend, title, etc. You may have to order the commands appropriately so all orbits appear on the figure. Plot Mercury in blue, Venus in yellow, Earth in green, and Mars in red.
- **(b)** Plot the orbits of Earth, Neptune, and Pluto in one figure using `polarplot` on the polar plane, making sure to have a legend, title, etc. As above, you may have to order the commands appropriately so all orbits appear on the figure. Plot Earth in green, Neptune in cyan and Pluto in magenta.

13. This problem will show the value of working with log-scales in plots. The efficacy of two drugs are given in Table 3.4 by measuring the percentage of binding of the

Table 3.4 Drug efficacy data.

Concentration level (M)	Binding of drug A (%)	Binding of drug B (%)
2×10^{-9}	2	0
5×10^{-9}	5	1
2×10^{-8}	19	4
7×10^{-8}	42	13
2×10^{-7}	68	31
7×10^{-7}	87	57
2×10^{-6}	95	80
9×10^{-6}	99	95
3×10^{-5}	100	98
5×10^{-4}	100	100

drug to the necessary receptor at different concentration levels in moles per liter, or molars (M).

(a) Plot the data points with concentration level of the drug on the horizontal axis and binding percentage on the vertical scale. First use a linear scale (using `plot`), being sure to label the axes and create a legend. Use different markers for drug A and drug B and connect the markers with a different line (solid, dotted, etc.).

(b) Create a similar plot of the data but instead use a semi-log scale (using `semilogx`).

CHAPTER 4

Three-Dimensional Plots

4.1. Vector functions or space curves

Vector functions (3D parametric equations) are defined and plotted similarly to 2D plots, except one must use the command `fplot3` instead of `fplot` and `plot3` instead of `plot`.

```
fplot3(@(t) cos(7*t), @(t) sin(3*t), @(t) cos(11*t))
```

One can fine-tune the domain used for `t` and color(s) used, but for more flexibility and control use `plot3`.

Step 1: Define your inputs (the t). This is where `linspace` comes in handy; make sure you SUPPRESS THE OUTPUT.

```
t = linspace(0,10*pi,200);
```

Step 2: Calculate the x, y, and z for each t value USING COMPONENT-WISE CALCULATIONS.

```
x = t.*cos(t);
y = 4*t+1;
z = t.*sin(t);
```

Step 3: Plot the defined vectors using the `plot3` command to get Fig. 4.1(A).

```
plot3(x,y,z)
xlabel('x'),ylabel('y'),zlabel('z')
```

As discussed in the previous chapter, you must be careful on how you define your domain to make sure you get the expected plot, as seen in Fig. 4.1(B).

```
% bad domain example
t = 0:10*pi;
x = t.*cos(t);
y = 4*t+1;
z = t.*sin(t);
plot3(x,y,z)
```

You can add plots, titles, label axes, etc. to your figure just as in 2D plots.

```
% multiple plots
```

Programming Mathematics Using MATLAB®
https://doi.org/10.1016/B978-0-12-817799-0.00009-0

```
t = linspace(0,3*pi);
x = t.*cos(t);
y = 4*t+1;
z = t.*sin(t);
s=linspace(-5,5);
x2 = 2*pi + s;
y2 = 2*pi*s;
z2 = 8*pi + 4*s;
plot3(x,y,z,x2,y2,z2)
xlabel('x'),ylabel('y'),zlabel('z')
```

You can add plots using `hold on` and `hold off`, and here we are also adding a point to the graph (see Fig. 4.2).

```
% multiple plots using hold on/hold off
t = linspace(0,10*pi,150);
x = t.*cos(t);
y = t.*sin(t);
z = 4*t;
s=linspace(-5,5);
x2 = 2*pi + s;
y2 = 2*pi*s;
z2 = 8*pi + 4*s;
plot3(x,y,z)
hold on
plot3(x2,y2,z2,'k')
xlabel('x'),ylabel('y'),zlabel('z')
plot3(2*pi,0,8*pi,'*k')      % plotting a point
hold off
```

As in 2D, you can change the window using `xlim`, `ylim` and/or `zlim` commands (try in the command window) or the `axis` command (see Fig. 4.3). Notice these commands

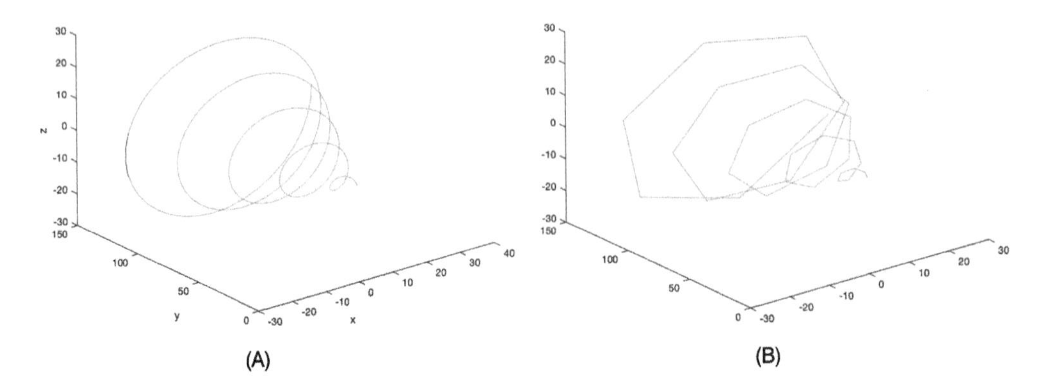

Figure 4.1 Vector functions. (A) Vector function example, (B) Bad domain example.

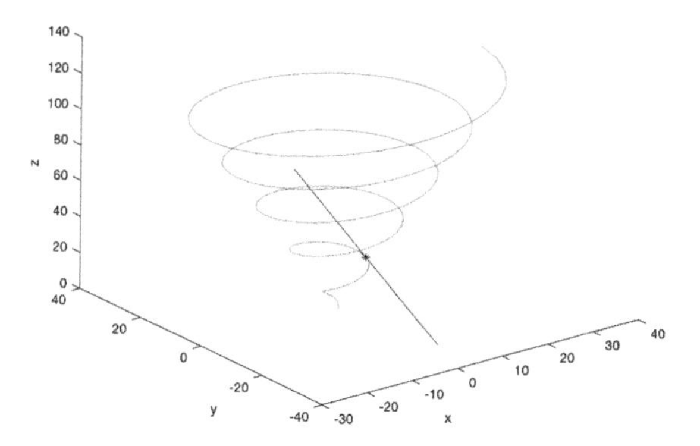

Figure 4.2 Multiple 3D plots.

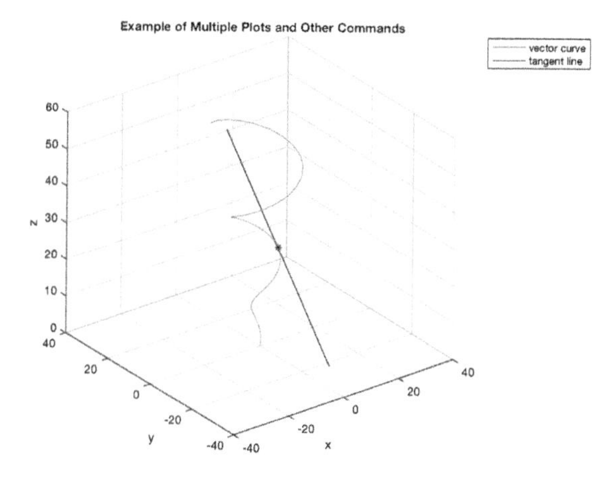

Figure 4.3 Multiple 3D plots.

may not get the desired result and it may be better to adjust your domains instead. The `grid on` command is also used in Fig. 4.3.

```
t = linspace(0,10*pi);
x = t.*cos(t);
y = t.*sin(t);
z = 4*t;
s=linspace(-5,5);
x2 = 2*pi + s;
y2 = 2*pi*s;
z2 = 8*pi + 4*s;
plot3(x,y,z)
hold on
plot3(x2,y2,z2,'k')
xlabel('x'),ylabel('y'),zlabel('z')
```

```
plot3(2*pi,0,8*pi,'*k')
zlim([0,60])
title('Example of Multiple Plots and Other Commands')
legend('vector curve', 'tangent line')
grid on
hold off %% DON'T FORGET TO HOLD OFF!!
```

4.2. Plotting surfaces

As for previous plots, you can use `fmesh` and/or `fsurf`.

```
fmesh(@(x,y) 0.5*cos(0.5*x).*sin(y))
%%
fsurf(@(x,y) exp(y).*cos(pi*x), [-4,4,0,2])
```

The commands that plot the three-dimensional surfaces establish points (x, y, z) of the surface to be corresponding elements in the matrices X, Y, and Z and then "connect-the-dots" as in the two-dimensional plots. One can create these matrices separately, but if they are not formed from datasets then many times the surface(s) are the result of functions. Thus the steps to establish the domain and then using component-wise calculations are especially important.

Example 4.2.1. Graph the surface $f(x, y) = \dfrac{4y}{x^2 + y^2 + 1}$ for $-10 \leq x \leq 10$ and $-5 \leq y \leq 5$.

Step 1: Establish the domain by creating vectors. They do not need to be the same size.

Step 2: Create the matrices X and Y based on the domain. The easiest way to do this is by using the command `meshgrid`.

Step 3: Calculate the corresponding Z using component-wise calculations on the matrices X and Y.

Step 4: Plot the surface. The most common commands are `mesh` and `surf` as seen in Fig. 4.4.

```
u = linspace(-10,10,50); v = linspace(-5,5,50);     % Step 1
[x,y] = meshgrid(u,v);                              % Step 2
z = (4*y)./(x.^2 + y.^2 + 1);                       % Step 3
mesh(x,y,z)                                         % Step 4
```

Note that Step 1 and Step 2 can be combined into one step, but for clarity they are shown separately.

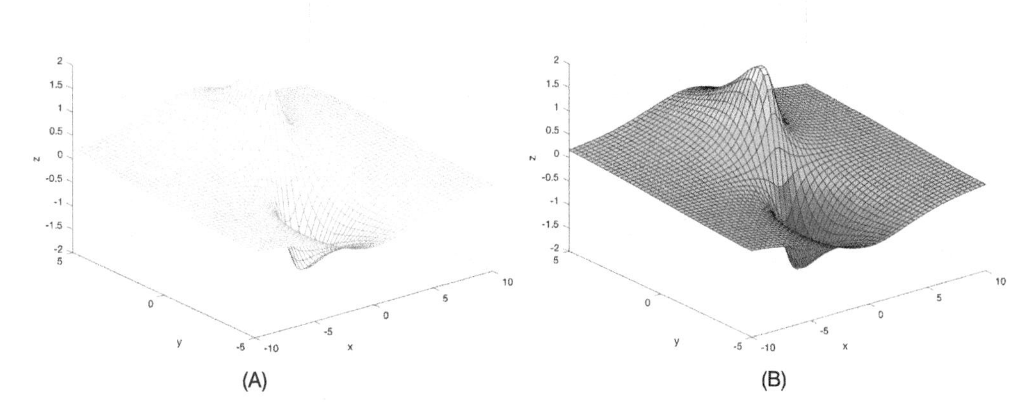

Figure 4.4 Plotting surfaces. (A) Using `mesh(x,y,z)`, (B) Using `surf(x,y,z)`.

4.2.1 The `meshgrid` command

The `meshgrid` command takes as inputs vectors to define matrices corresponding matrices. Once can create two or three matrices. For the purpose of this example, we will create two matrices. Note that the vectors for the domains below are not the same size, and are created within the `meshgrid` command.

```
>> [x,y]=meshgrid(1:5,6:8)
>> z=x.*y
x =
     1     2     3     4     5
     1     2     3     4     5
     1     2     3     4     5
y =
     6     6     6     6     6
     7     7     7     7     7
     8     8     8     8     8
z =
     6    12    18    24    30
     7    14    21    28    35
     8    16    24    32    40
```

Once the matrices are created, three-dimensional plot commands will use as points the corresponding elements of these matrices, and then connect those points. Thus the first point plotted in the above example will be the point $(x, y, z) = (1, 6, 6)$, and the last point will be $(5, 8, 40)$ (see Fig. 4.5).

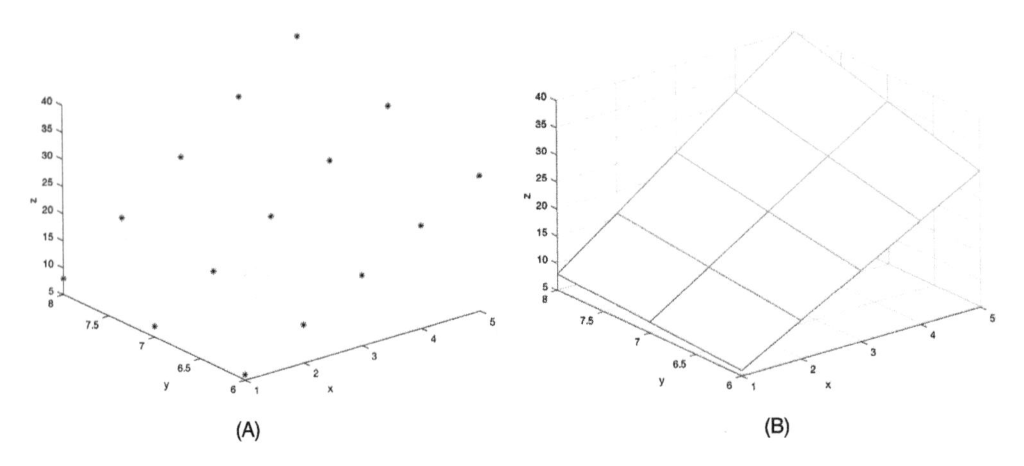

Figure 4.5 `meshgrid` examples. (A) `plot3(x,y,z,'*k')`, (B) `mesh(x,y,z)`.

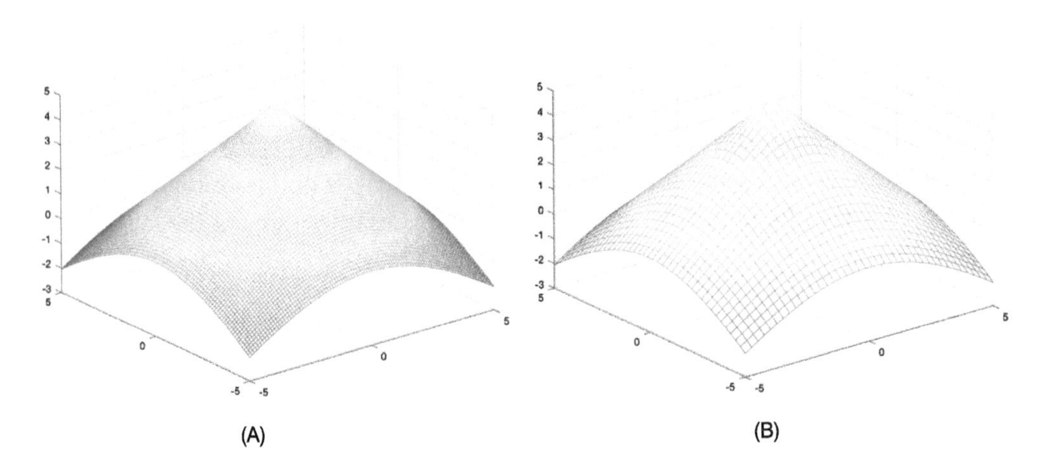

Figure 4.6 Domain examples. (A) Default `linspace`, (B) Using 40 values.

4.2.2 Domain issues

There can be a fine line between not enough and too many points for creating accurate, aesthetically pleasing figures. You may notice that in the above example to create Fig. 4.4, the command `linspace` was used to create the vectors but the number of values was decreased from the default 100 to 50. Using the default can create a surface that may not be aesthetically pleasing and/or may slow MATLAB down by performing more calculations than necessary to create the surface (see Fig. 4.6).

```
[x,y]=meshgrid(linspace(-5,5));
z=5-sqrt(x.^2 + y.^2);
surf(x,y,z)
```

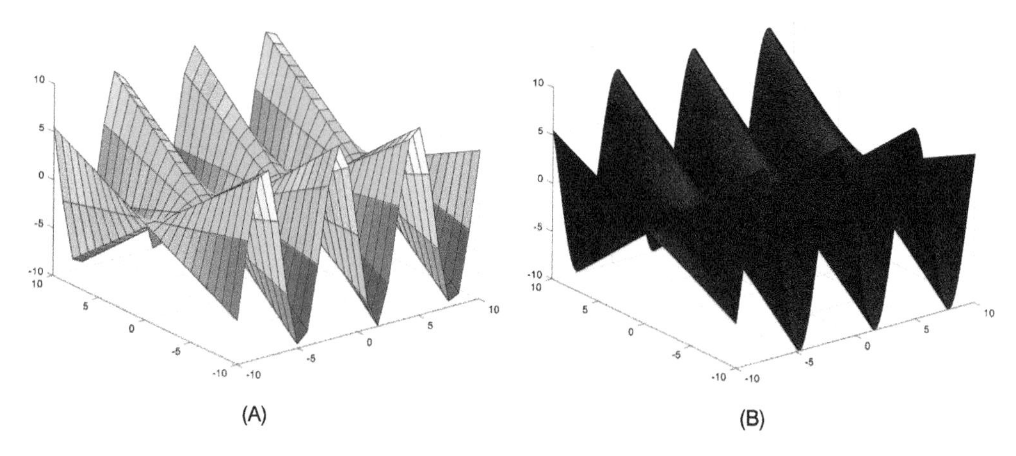

Figure 4.7 Bad domain examples. (A) `linspace(-9,9,25)`, (B) `[x,y]=meshgrid(-9:.05:9)`.

Notice in the code above, the `meshgrid` command has one input vector. This command is equivalent to the command `meshgrid(linspace(-5,5),linspace(-5,5))` but is more efficient.

If one does not use enough values, the surface can appear jagged, as in Fig. 4.7(A). Too many values as in Fig. 4.7(B) can create a bad figure, also. Note that for some surfaces even having 40 instead of 60 in your vector can make it appear jagged. This second example shown in Fig. 4.7(B) was created by using the command `[x,y]=meshgrid(-9:.05:9);` for `z=y.*sin(x)`. This is common if one is in the habit of using the colon operator for creating domain vectors in 2D plots. When using this notation to create 3D plots, this can create vectors and thus matrices of a much larger size than needed, and can greatly slow down MATLAB in performing both the calculations command(s) and plotting command(s). Think about the size of the matrices created if the domain was from -100 to 100 and one created a vector `-100:.01:100`!

4.2.3 Level curves

In multivariable calculus and other disciplines, level curves are useful to study functions of two variables or three-dimensional surfaces. In MATLAB®, level curves are contour plots. One sets up these as one does with surfaces, but uses commands such as `contour`, `contourf` or `contour3`.

Example 4.2.2. Plot the level curves of $f(x, y) = -xye^{-x^2-y^2}$.

```
[x,y]=meshgrid(linspace(-4,4));
z = -150*x.*y.*exp(-x.^2-y.^2);
contour(x,y,z)
```

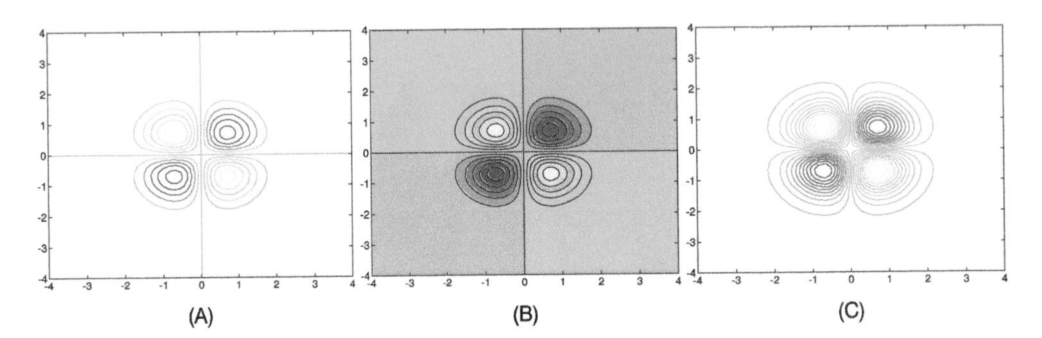

Figure 4.8 Level Curves. (A) `contour(x,y,z)`, (B) `contourf(x,y,z)` and (C) `contour(x,y,z,20)`.

The above code creates Fig. 4.8(A). You can create a filled contour as in Fig. 4.8(B) using `contourf`.

One can adjust the number of level curves shown by having a fourth argument as in Fig. 4.8(C); otherwise MATLAB automatically chooses the number of level curves drawn.

To create labels, or to adjust which values the level curves are shown; see the code below. The output of the code below is in Fig. 4.9.

```
[c,h]=contour(x,y,z);
clabel(c,h)
%% Forcing values of levels (with labels)
cvalues=-20:4:20;
[c,h]=contour(x,y,z,cvalues);
clabel(c,h)
```

You can show level curves in 3D, and even add a mesh surface or filled surface as in Fig. 4.10.

4.2.4 Multiple plots and modifying colors

Using `hold on` and `hold off`, one can add multiple plots to 3D graphs as in the 2D case. One can also change the colors in the `mesh` and `surf` plots by setting `EdgeColor` and/or `FaceColor` as in Fig. 4.11. One can also combine surfaces, vector functions and/or points within one figure with the `hold on` and `hold off` commands. Common mistakes are to forget to use `plot3` rather than `plot` in these instances, or to use `plot3` when `mesh` or `surf` is more appropriate.

```
[x,y]=meshgrid(linspace(-2,2,55));
z1=y.*exp(x.^2-5);
```

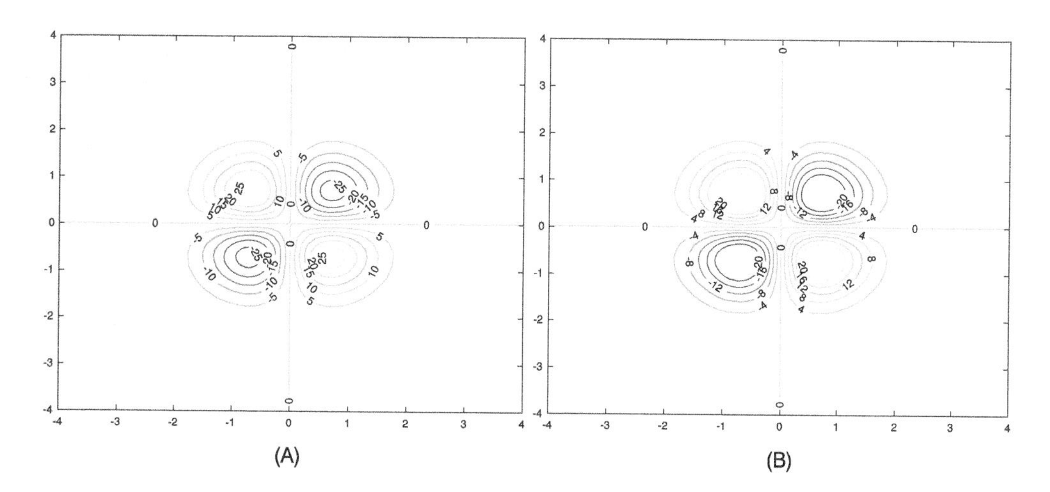

Figure 4.9 Modifying level curves. (A) Level Curve with Labels, (B) Specifying Levels.

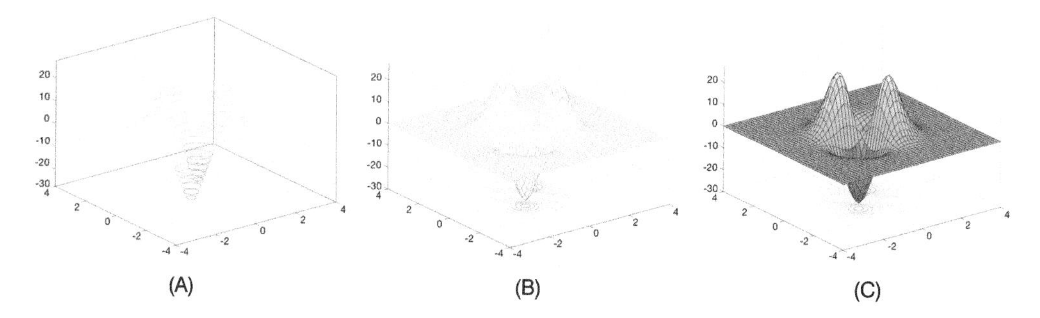

Figure 4.10 Level curves in 3D. (A) `contour3(x,y,z,20)`, (B) `meshc(x,y,z)` and (C) `surfc (x,y,z)`.

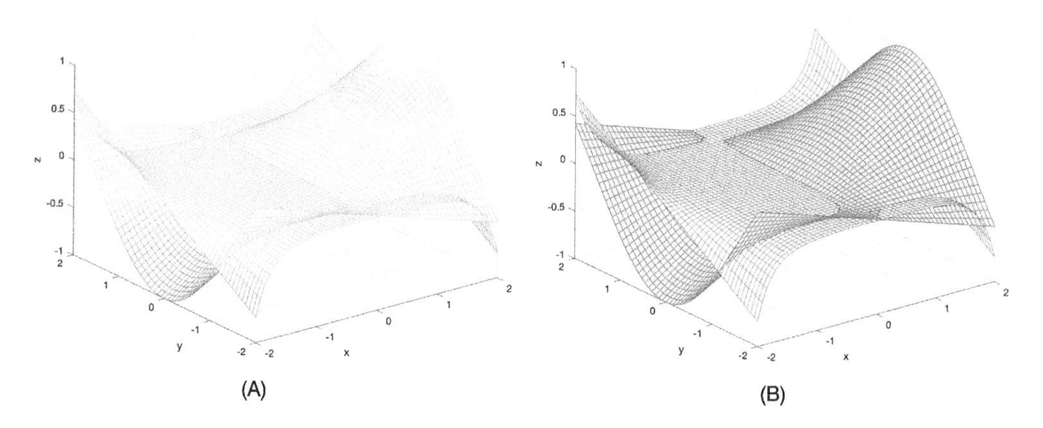

Figure 4.11 Multiple plots and modifying colors. (A) No modifications, (B) Specifying Colors.

```
mesh(x,y,z1)
xlabel('x'),ylabel('y'),zlabel('z')
hold on
z2=1/2*x.*cos(y);
mesh(x,y,z2)
hold off
%% Modifying colors
LoyGreen = 1/255*[0, 104, 87];
mesh(x,y,z1,'EdgeColor',LoyGreen)
xlabel('x'),ylabel('y'),zlabel('z')
hold on
surf(x,y,z2,'FaceColor','y', 'EdgeColor','r')
hold off
```

The above code above creates Figs. 4.11(A) and 4.11(B).

The default "colormap" of 3D plots changed in R2014b. Previous to this edition of MATLAB, the default colormap was "jet." It is now "parula" (see Fig. 4.12). There are many other pre-set colormaps, and you can create your own. See MATLAB's documentation on colormap for a full treatment.

4.3. View command

Within the MATLAB figure window, you may have found the "Rotate 3D" capability quite useful. If instead, you want a static picture to have a certain perspective in 3D, you can use the `view` command. When you rotate the 3D figure window, you may notice some numbers appearing in the bottom left corner of the window (see Fig. 4.13).

The azimuth is the angle measured from the negative y-axis in the positive (counterclockwise) direction on the xy-plane. The elevation is the angle from the xy-plane towards the positive z-axis. These angles (measured in degrees) determine the viewpoint looking at the plot, as shown in Fig. 4.14.

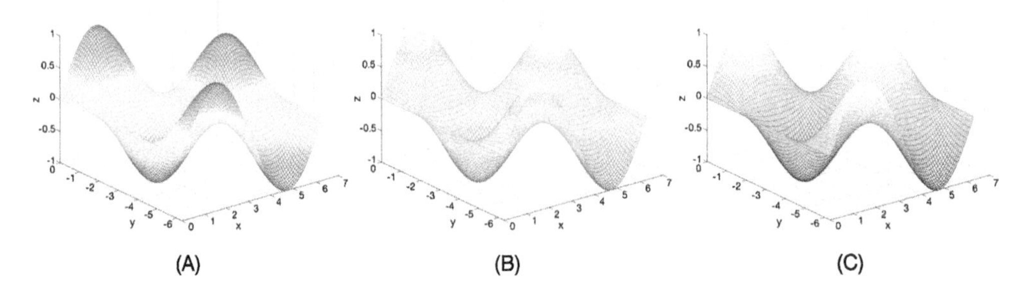

Figure 4.12 Colormaps. (A) `colormap jet`, (B) `colormap default` and (C) `colormap bone`.

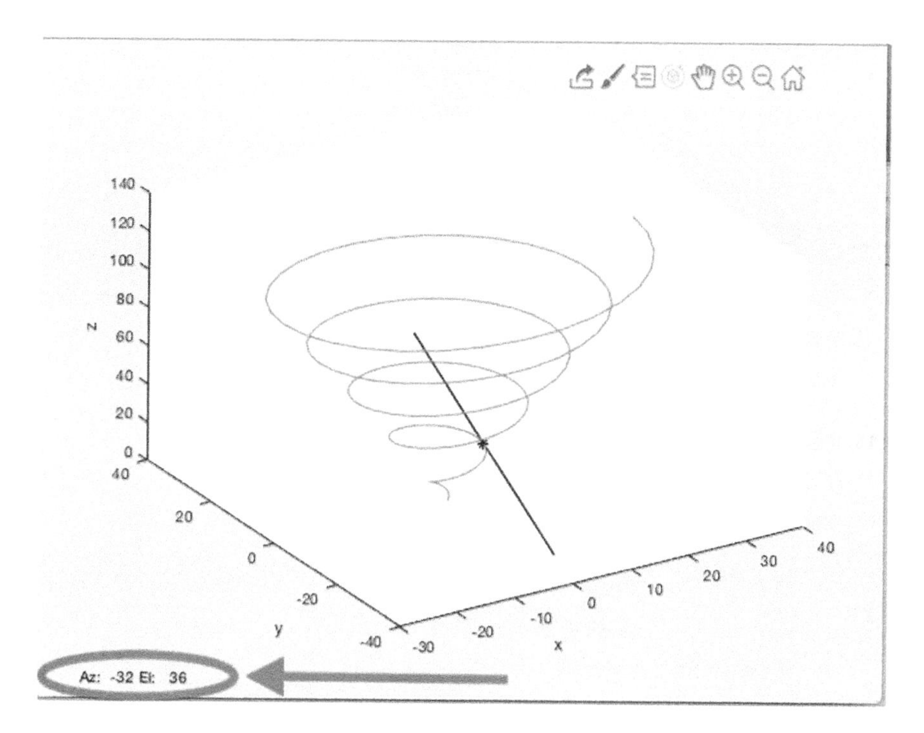

Figure 4.13 Rotate 3D screen shot.

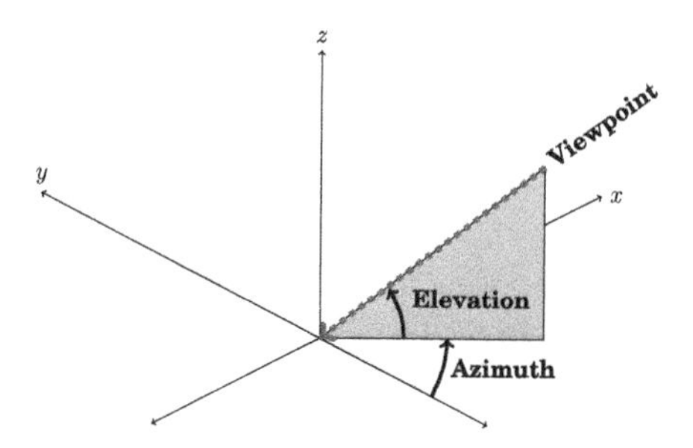

Figure 4.14 Explanation of `view`.

The default view is `view(-37.5, 30)`, as seen in Fig. 4.15(A). By changing the viewpoint to be similar to what is shown in Fig. 4.14 (using `view(30,30)`), we get Fig. 4.15(B).

Using the `view` command, you can look down at the *xy*-plane, *xz*-plane, and *yz*-plane (see Fig. 4.16). This will be explored in the exercises.

As mentioned in the MATLAB documentation, there are some limitations of control with this command. To gain more control of the view, you can use the camera properties but these are not discussed here.

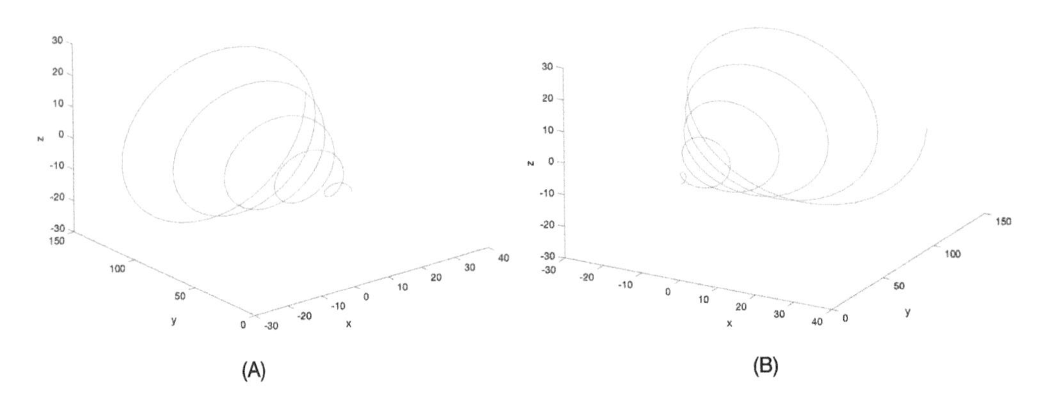

Figure 4.15 View examples. (A) The default view, (B) Using `view(30,30)`.

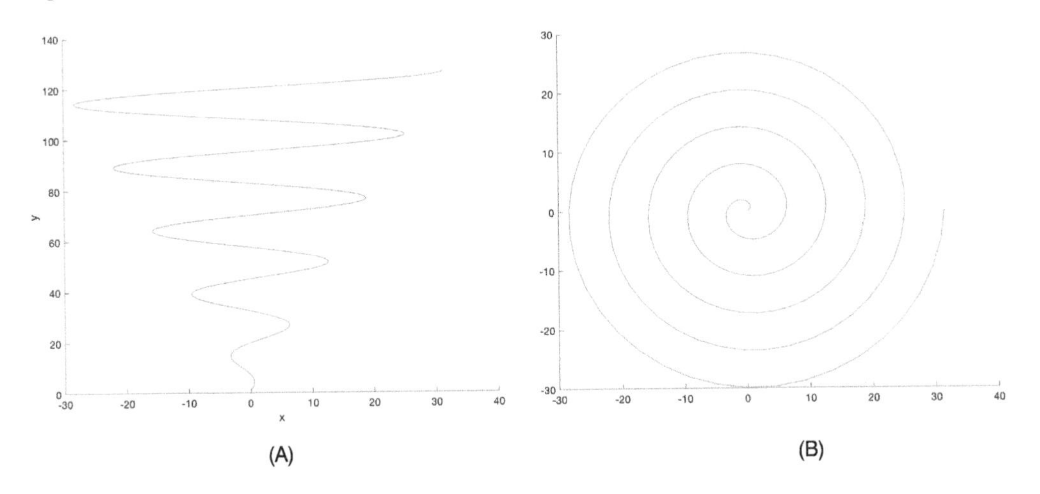

Figure 4.16 View examples. (A) *xy*-plane view, (B) *xz*-plane view.

4.4. Axis settings, revisited

As discussed in Section 3.3, the axis settings can really change the look of the plots. These settings can be especially important and especially tricky in 3D graphs, as the next example demonstrates.

Example 4.4.1. Graph the surface $f(x, y) = 2x^2 + y^2$ along with its tangent plane and normal vector at the point $(1, 1, 3)$.

From multivariable calculus, we get that the equation for the tangent plane is $z = 4x + 2y - 3$ and the parametric equations for the normal vector are

$$x = 1 + 4t,$$

$$y = 1 + 2t,$$

$$z = 3 - t.$$

Graphing it and then adjusting the view gives us the figures in Fig. 4.17.

```
% surface
[x,y]=meshgrid(linspace(-4,4,50));
z=2*x.^2+y.^2;
% plane
z2 = 4*x+2*y-3;
% vector
t=linspace(-1,1.5);
x3=1+4*t;
y3=1+2*t;
z3=3-t;
% graphing
mesh(x,y,z,'EdgeColor','black')
hold on
surf(x,y,z2,'FaceColor','blue')
plot3(x3,y3,z3,'r', 'LineWidth',2)
hold off
xlabel('x'),ylabel('y'),zlabel('z')
```

When first seeing these figures, you may check and recheck their calculus and wonder what is going on as the normal vector does not appear to be perpendicular to the tangent plane. The problem is not the calculus; it is the aspect ratios of these figures. These can be fixed by using `axis equal` with some adjustments. If using `zlim` still does

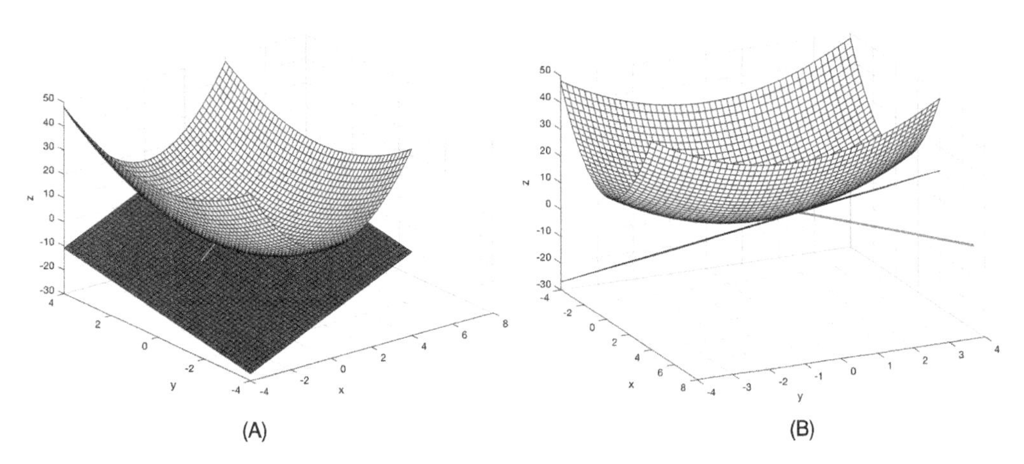

Figure 4.17 Different `view` settings for Example 4.4.1. (A) default view: `view(-37.5,30)`, (B) `view(65,25)`.

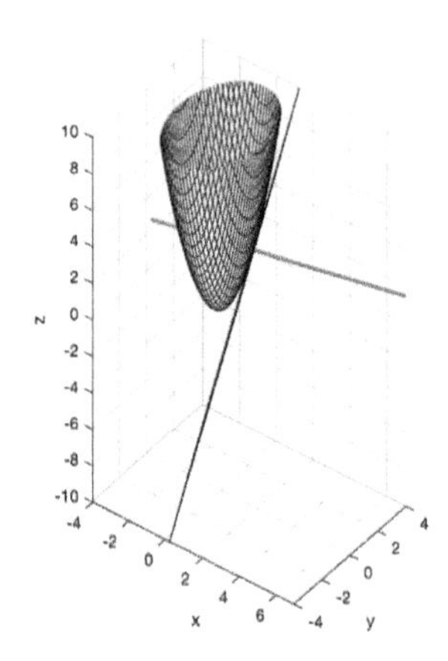

Figure 4.18 `axis equal, view(37,40), zlim([-10,10])`.

not achieve the desired picture, as mentioned previously you may also want to adjust the original domains to get the desired picture (see Fig. 4.18).

4.5. Other coordinate systems and 3D graphs

4.5.1 The `sphere` and `cylinder` commands

The `sphere` and `cylinder` commands are quite useful commands.

To plot a unit sphere, use the commands `sphere` and `axis equal`. As discussed above, without the `axis equal` command it will not look like a perfect sphere. One can also force more or less faces to create the sphere by using the command `sphere(n)`, where n is the number of desired faces (see Fig. 4.19). The default is 20, thus `sphere` is equivalent to `sphere(20)`.

The most flexibility comes from using the sphere command in the form of `[x,y,z]=sphere(n)`. This generates three $(n+1)\times(n+1)$ matrices `x`, `y`, and `z` that when combined with the `mesh(x,y,z)` or `surf(x,y,z)` commands, will generate a sphere with n faces. This allows you to translate, resize, and/or recolor the sphere as you would any 3D graph (see Fig. 4.20).

```
[x,y,z]=sphere(30);
mesh(x,y,z)                        % unit sphere
hold on
surf(x+3,y+2,z-1)       % sphere with center (3,2,-1)
r=0.5;
mesh(r*x-1,r*y+2,r*z+1,'FaceColor',[0.5,0.5,0.5],...
 'EdgeColor', 'k')      % sphere resized, moved and recolored
axis equal
xlabel('x'),ylabel('y'),zlabel('z')
```

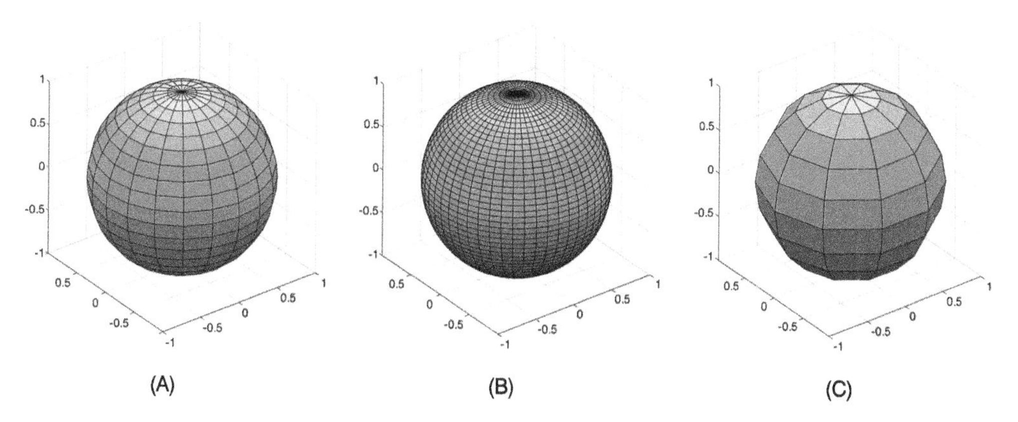

Figure 4.19 Spheres using `axis equal`. (A) default `sphere`, (B) `sphere(50)` and (C) `sphere (10)`.

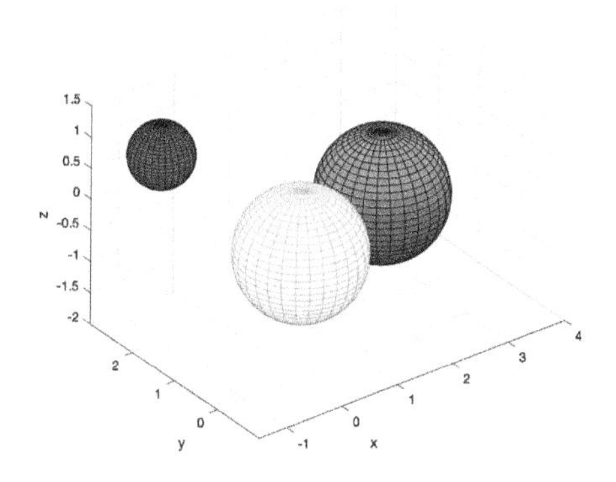

Figure 4.20 Drawing different spheres with `[x,y,z]=sphere(n)`.

The command `cylinder` works similarly to the `sphere` command and will generate a unit circular cylinder: a cylinder of radius 1 and height 1 with the center of circle at $(0, 0)$ in the xy-plane (see Fig. 4.21(A)). The command `cylinder(f)` will create a cylinder using `f` as the profile curve; in other words, `f` is the radius of the cylinder at equally spaced heights along the cylinder (see Fig. 4.21(B)). Thus `f` could be one value of a vector of values. The command `cylinder` is equivalent to `cylinder(1)`. The command `cylinder(f,n)` will create a cylinder using `f` as the profile curve but using `n` equally spaced points around its circumference (see Fig. 4.21(C)). Thus `cylinder` is equivalent to `cylinder(1,20)`.

As in the case with the `sphere` command, using `[x,y,z]=cylinder(f,n)` allows for more control over modifying the cylinders generated using transformations on (x, y, z) (see Fig. 4.22).

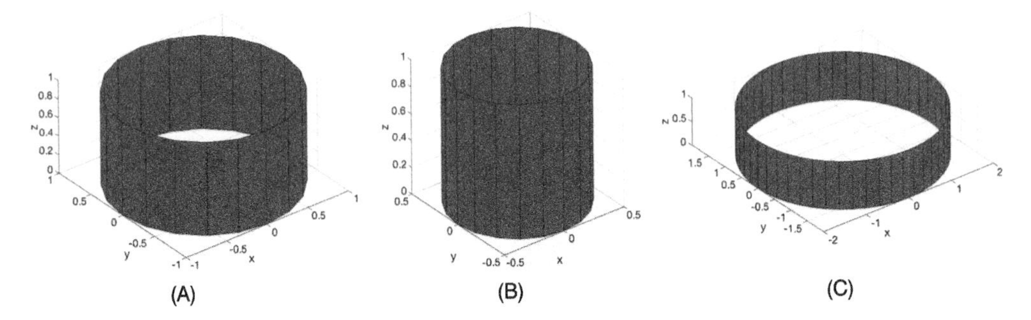

Figure 4.21 Cylinders using `axis equal`. (A) default `cylinder`, (B) `cylinder(0.5)` and (C) `cylinder(2,50)`.

```
[x,y,z]=cylinder(1,50);
mesh(x,y,z)
hold on
surf(x-1,y+2,2*z)
surf(2*x-1,2*y+1,0.5*z+3,'FaceColor',[0.3,0.3,0.3])
hold off
axis equal
xlabel('x'),ylabel('y'),zlabel('z')
```

As mentioned above, the value(s) of `f` can be a vector of values, or as in the case below determined by values of a function (see Fig. 4.23).

```
t=linspace(0,2*pi);
f=2*sin(t) + t + 2;
[x,y,z]=cylinder(f,50);
mesh(x,y,z)
xlabel('x'),ylabel('y'),zlabel('z')
%%
v1=[0,10];
v2=0:10;
v3=0:0.5:10;
v4=v3.^2;
[x,y,z]=cylinder(v2,50);
mesh(x,y,z)
xlabel('x'),ylabel('y'),zlabel('z')
```

Experiment with the above code to see the difference between using `v1`, `v2` (shown), `v3`, and `v4`.

If you would like to have the cylinder horizontal instead, it may be easiest to define the matrices as usual but switch the variables in the plotting command (see Fig. 4.24). You can then adjust the cylinder's length, width, and center via transformations similar to what is done in Fig. 4.22. These adjustments will be explored in the exercises.

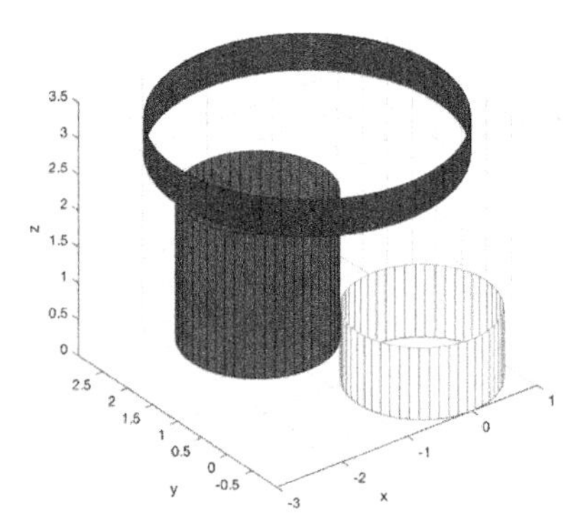

Figure 4.22 Drawing different cylinders with `[x,y,z]=cylinder(f,n)`.

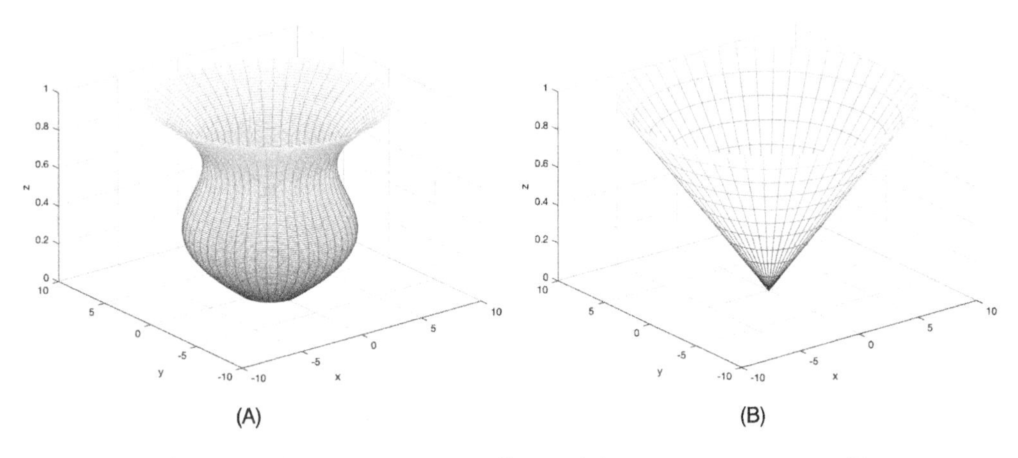

Figure 4.23 Cylinders with non-constant profile `f`. (A) `cylinder(f,50)`, (B) `cylinder (0:10,50)`.

```
t=linspace(0,2*pi);
f=2*sin(t)+t + 2;
[x,y,z]=cylinder(f,50);
mesh(3*z+2,x,y)      % horizontal, shifted cylinder
xlabel('x'),ylabel('y'),zlabel('z')
```

4.5.2 Cylindrical coordinates

To plot a function given in cylindrical coordinates, one can always use the commands to convert to Cartesion coordinates after using the `meshgrid` command on θ and r or one can use the `[x,y,z]=pol2cart(theta,r,z)` command. Both ways are shown in Example 4.5.1 and in Fig. 4.25.

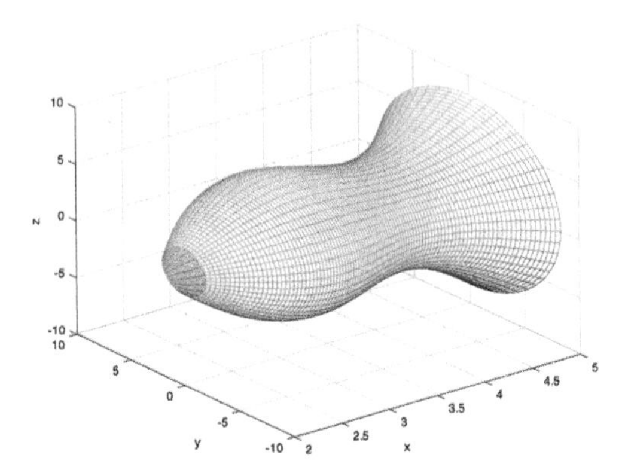

Figure 4.24 Creating a horizontal cylinder.

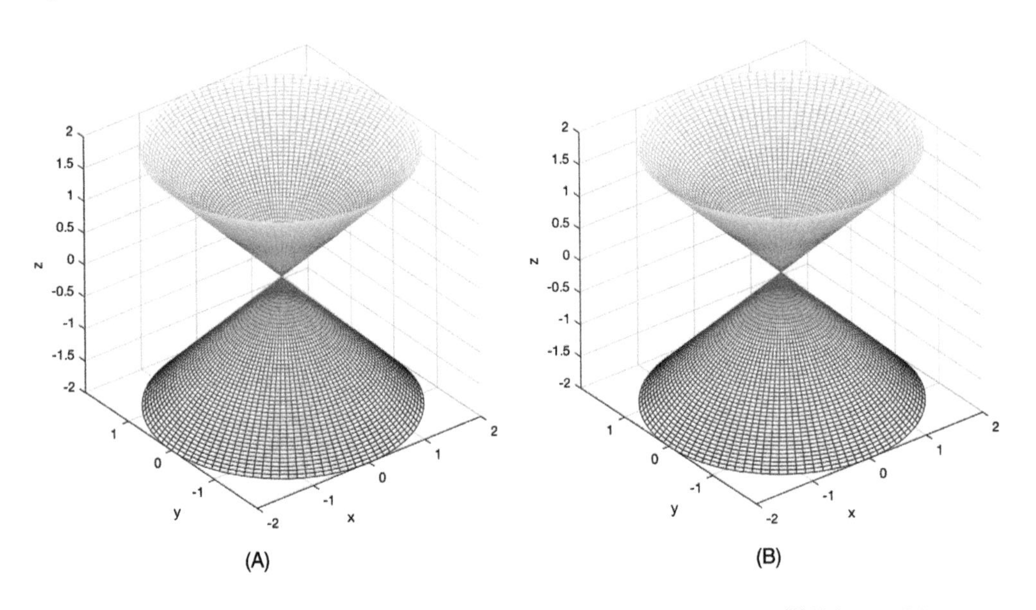

Figure 4.25 Cylindrical coordinates example. (A) Using conversion equations, (B) Using `pol2cart`.

Example 4.5.1. Graph the surface $z = r$ that is given in cylindrical coordinates for $r \in [-2, 2]$ and $\theta \in [0, 2\pi]$.

```
%% Using Conversion Equations
tdomain=linspace(0,2*pi); rdomain = linspace(-2,2);
[t,r]=meshgrid(tdomain,rdomain);
x = r.*cos(t);
y = r.*sin(t);
z = r;
mesh(x,y,z)
```

```
xlabel('x'),ylabel('y'),zlabel('z')
axis equal
%% Using pol2cart
tdomain2=linspace(0,2*pi); rdomain2 = linspace(-2,2);
[t2,r2]=meshgrid(tdomain2,rdomain2);
z2=r2;
[x2,y2,z2]=pol2cart(t2,r2,z2);
mesh(x2,y2,z2)
xlabel('x'),ylabel('y'),zlabel('z')
axis equal
```

4.5.3 Spherical coordinates

Using spherical coordinates can be useful; unfortunately the definitions of θ and ϕ differ from many calculus texts. In some calculus texts you have the triple (r, θ, ϕ) where θ is the azimuth angle as with the `view` command and ϕ is the zenith angle, or the angle measured from the positive z-axis. In physics, the definitions of θ and ϕ are reversed. In MATLAB, θ is defined as the azimuth angle but ϕ is measured as the angle of elevation as in the `view` command discussed in Section 4.3 and shown in Fig. 4.14.

As with cylindrical coordinates, one may use the conversion equations or the command `sph2cart`, keeping in mind the definition MATLAB has for ϕ (see Fig. 4.26).

```
%% using calculus definition of phi
% and usual conversion equations
theta=linspace(0,2*pi);
phi=linspace(0,pi/4);
[t,p]=meshgrid(theta, phi);
rho = 3+0*t;  % the 0*t ensures rho is correct size
x = rho.*cos(t).*sin(p);
y = rho.*sin(t).*sin(p);
z = rho.*cos(p);
mesh(x,y,z)
xlabel('x'), ylabel('y'), zlabel('z')
axis equal
%% using MATLAB's definition of phi
% and the adjusted conversions
clf
theta=linspace(0,2*pi);
phi=linspace(0,pi/4);
[t,p]=meshgrid(theta, phi);
rho = 3+0*t;
x = rho.*cos(t).*cos(p);
y = rho.*sin(t).*cos(p);
z = rho.*sin(p);
mesh(x,y,z)
xlabel('x'), ylabel('y'), zlabel('z')
```

```
axis equal
%% using sph2cart
clf
theta=linspace(0,2*pi);
phi=linspace(0,pi/4);
[t,p]=meshgrid(theta, phi);
rho = 3+0*t;
[x,y,z]=sph2cart(t,p,rho);
mesh(x,y,z)
xlabel('x'), ylabel('y'), zlabel('z')
axis equal
```

Note that there are other commands such as `cart2sph` and `cart2pol` but these are not discussed here.

4.6. Exercises

1. A helix is a three-dimensional curve (shape of a spring) that can be modeled by the following equations:

$$x = a\cos t,$$
$$y = a\sin t,$$
$$z = bt,$$

where a is the radius of the helical path and b is a constant that determines the "tightness" of the path.

(a) Plot a basic helix ($a = b = 1$) for five complete turns (be careful: what should be your domain for t equal)?

The DNA molecule has the shape of a double helix, where one helix can be modeled by the above equations. The radius of each helix of the DNA molecule is about 10 angstroms (1 Å $= 10^{-8}$ cm). Each helix rises about 34 Å during each complete turn [26].

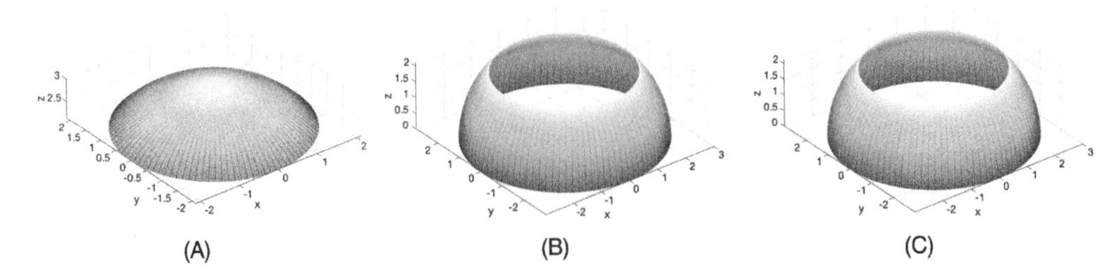

Figure 4.26 Spherical coordinates. (A) Calculus definition of ϕ, (B) MATLAB definition of ϕ and (C) Using `sph2cart`.

(b) Figure out what a and b should equal for the single DNA helix. Work on paper (if any) should be turned in and the answers should be text on the webpage. Have the units in angstroms and state the EXACT VALUES.

(c) Plot five complete turns of the DNA helix (what should be the domain for t equal?).

(d) Plot the helix for 100 complete turns. BE CAREFUL! In order for this to be shown correctly (i.e., showing all of the turns), you will have to define your domain correctly.

(e) Estimate the full length of ONE COMPLETE TURN of the DNA helix using the arc length formula

$$L = \int_c^d \sqrt{\left(\frac{dx}{dt}\right)^2 + \left(\frac{dy}{dt}\right)^2 + \left(\frac{dz}{dt}\right)^2}\, dt.$$

Figure out what c and d should be in the integral and calculate L on paper using calculus, showing all work. Turn in your work on paper and state the EXACT VALUE of your answer as text on the webpage (in angstroms). Also use MATLAB to give a numerical approximation of this answer in angstroms.

(f) In one cell in the human body, it has been said that DNA makes about 2.9×10^8 complete turns. Using your (exact value) answer in part (1e), use MATLAB to give a numerical approximation of the FULL LENTH of the DNA helix, **in meters**.

(g) It has been said that there are about 10^{13} cells in the human body. Use this, along with your answer in part (1f) to estimate the length of all of the DNA in a human body. The average distance between the Earth and Mars is 225 million km [22]. Would the stretched out DNA in a human body reach Mars? If so, how many roundtrips would there be from the Earth to Mars and back?

2. Graph the following 3D surfaces. For each of these, have the titles specify the problem number and part, and if the surface has a special name include that as well. Label the axes.

(a) $z = 1 - |x + y| - |y - x|$. Use `fsurf`.

(b) $z = x^3 - 3xy^2$, $x, y \in [-15, 15] \times [-15, 15]$ (monkey saddle). Use `mesh`.

(c) $z = \dfrac{\sin(x^2 + y^2)}{(x^2 + y^2)}$, $x, y \in [-5, 5] \times [-5, 5]$. Use `surf` and have the z-axis go from -2 to 2.

(d) $z = \operatorname{sgn}(xy)\operatorname{sgn}(1 - 100x^2 + 100y^2)$, $x, y \in [-10, 10] \times [-10, 10]$. Use `surf` and 50 points each for the domains of x and y.

(e) $z = \operatorname{sgn}(xy)\operatorname{sgn}(1 - 100x^2 + 100y^2)$, $x, y \in [-10, 10] \times [-10, 10]$. Use `mesh` and 200 points each for the domains of x and y.

(f) $z = \sin x + \sin y$, $x, y \in [0, 6\pi] \times [0, 20]$. Use `axis equal` and your choice of `mesh` or `surf`.

(g) Graph the level curves (contour lines) for $z = \sin(x - y)$, $x, y \in [-10, 10] \times [-10, 10]$.

(h) Graph the level curves (contour lines) for $z = 1 - |x + y| - |y - x|$, $x, y \in [-15, 15] \times [-15, 15]$.

3. Graph the following 3D surfaces. For each of these, have the titles specify the problem number and part, and have the axes labeled.

(a) $z = x^3y - xy^3$, $x, y \in [-5, 5] \times [-5, 5]$ (dog saddle). Use `surf`.

(b) $z = \cos(x - y)$, $x, y \in [-2\pi, 2\pi] \times [-2\pi, 2\pi]$. Use `mesh`.

(c) $z = e^x \cos(y)$, $x \in [0, 3]$, $y \in [-3\pi, 3\pi]$. Your choice of `mesh` or `surf`.

(d) $z = \cos x - \cos y$, $x, y \in [-10, 10] \times [-10, 10]$ our choice of `mesh` or `surf`.

(e) Graph the level curves (contour lines) for $z = \cos(x - y)$, $x, y \in [-10, 10] \times [-10, 10]$.

4. Consider the 3D parametric surface that is known as the Möbius strip.

$$\mathbf{m}(u, v) = \big((7 + v\cos(u))\cos(2u),\ (7 + v\cos(u))\sin(2u),\ (v\sin(u))\big)$$

for $(u, v) \in [0,\ 2\pi] \times [-1, 1]$. All graphs should have their domains defined appropriately so there are enough points, but not too many points. All axes should be labeled appropriately.

(a) Graph the strip $\mathbf{m}$ using `mesh`. Make the z-axis from -2 to 2.

(b) Note that if we fix one of the variables u or v, the result is a vector function or space curve. Add graphs of the "edges" of the strip by creating two space curves; one when $v = 1$ and the other when $v = -1$. Plot the first curve black and the second in white. How many edges does the strip have?

(c) Add graphs of space curves when $v = 0$ and $v = 0.5$, the first as blue and the second as red.

(d) Graph just the space curves from parts (b) and (c).

(e) Consider a general Möbius strip

$$\mathbf{m}(u, v) = \big((a + bv\cos(u))\cos(2u),\ (a + bv\cos(u))\sin(2u),\ (bv\sin(u))\big)$$

for $(u, v) \in [0,\ 2\pi] \times [-1, 1]$. The strip above is when $a = 7$ and $b = 1$. Graph the strip for different values of a and b. What changes about the strip when we change those values? What happens if we change the domain for v to $[-5, 5]$?

5. Graph the following 3D surfaces. For each of these, have the titles specify the problem number and part, and have the axes labeled.

(a) $z = \sin(x - y)$, $x, y \in [-\pi, \pi] \times [-2\pi, 2\pi]$. Use `surf` and the default number of elements when defining your domain.

(b) $z = \sin(x - y)$, $x, y \in [-\pi, \pi] \times [-2\pi, 2\pi]$. Use `surf` and when defining your domain, use fewer than the default to get a decent, yet not jagged, graph.
(c) Graph the level curves (contour lines) for $z = \sin(x - y)$, $x, y \in [-10, 10] \times [-10, 10]$.

6. Graph the function

$$f(x, y) = x\cos\left(\frac{2\pi x}{50}\right)\cos\left(\frac{2\pi y}{50}\right), \quad x \in [0, 150], y \in [0, 100].$$

Create the plots using `mesh`, `surf` and create a level curve (contour plot) of $f(x, y)$ and display all three of them side-by-side using `subplot`. Explain on paper what the difference between mesh and surf are, which you prefer, and why. Make sure each of the axes are labeled and each plot has an appropriate title.

7. Graph the function f and a contour plot of the function

$$f(x, y) = \sin\left(\frac{2\pi x}{60}\right)\sin\left(\frac{3\pi y}{60}\right)$$

in the domain $x \in [0, 100]$ and $y \in [0, 100]$. Graph the plots side-by-side.

8. Use the `sphere` and `surf` commands to create a sphere of radius 4, centered at the point $(-2, 3, 5)$. Also, make the sphere a color of your choice.
9. Use the `cylinder` and `mesh` commands to create a cylinder of radius 5, height 7 and with axis at the vertical line with $x = -2$ and $y = 3$.
10. Make the above cylinder horizontal.
11. Create a solid of revolution such as one would see in integral calculus. First, graph $f(x) = x|\sin(x)| + 4$ from $x = 1$ to $x = 8$. We will create a solid of revolution by rotating it about the x-axis. Similar to Fig. 4.24, use the `cylinder` command and `mesh` to create this solid so that it is horizontal and the x-values go from 1 to 8. Label the axes appropriately.
12. Consider the surface $xe^{yz} = 1$ at the point $(1, -1)$.
    (a) Find the tangent plane and normal line at that point.
    (b) Graph the surface, tangent plane, and normal line. For the surface and plane, the domains should be $x, y \in [0.5, 1.5] \times [-1.5, -0.5]$. The domain for the line should be $t \in [-1, 1]$. Make sure your domains are defined appropriately to be able to see the picture, and modify the aspect ratio to see the orthogonality of the line with the surface. Have the line in black, the surface $z = f(x, y)$ be one color, and the plane be another. The use of the command `alpha` and/or `view` may be useful to get a decent figure.

CHAPTER 5

Functions

5.1. The `lookfor` and `help` commands

The command `lookfor` is a useful command when you think there may be a command with a certain name, or that performs a certain function.

```
>> lookfor transpose
ctranspose                     - {\S}\textquotesingle{\S}   Complex conjugate
     transpose.
transpose                      - .{\S}\textquotesingle{\S} Transpose.
tfqmr                          - Transpose Free Quasi-Minimal Residual Method.
finargflip                     - Transpose array arguments to conform to size
     conventions.
tsAlignSizes                   - If the time vector is aligned to differing
     dimensions, a 'transpose' is
```

```
>> lookfor 'dot product'
dot                            - Vector dot product.
```

Once you know the command name, you can use the `help` command as discussed previously.

```
>> help dot
 dot  Vector dot product.
   C = dot(A,B) returns the scalar product of the vectors A and B.
   A and B must be vectors of the same length.  When A and B are both
   column vectors, dot(A,B) is the same as A'*B.

   dot(A,B), for N-D arrays A and B, returns the scalar product
   along the first non-singleton dimension of A and B. A and B must
   have the same size.

   dot(A,B,DIM) returns the scalar product of A and B in the
   dimension DIM.

   Class support for inputs A,B:
      float: double, single

   See also cross.

   Reference page in Help browser
      doc dot
```

Programming Mathematics Using MATLAB®
https://doi.org/10.1016/B978-0-12-817799-0.00010-7

MATLAB® `help` and `lookfor` commands will search both the native MATLAB functions but also functions within the current folder or directory.

5.2. File format

There are many reasons why one would want to write their own functions. MATLAB functions are M-files (extension `.m`) but with a certain format.

```
function header
% H1 LINE
% HELP LINE(S)
BODY
end % end line not always necessary, but useful for clarity
```

The first line of the file (header) MUST be of the form
`function outputvariable = functionname(inputvariable)`
To avoid confusion, the function name should be the name of the M-file (without the `.m`).

The H1 line is the line that is used with the `lookfor` command. The command searches the H1 lines for the keyword(s) given, and if found it prints the H1 line. Note that if the keyword is part of the function name, it will not show the function unless the function name appears on the H1 line.

For example, consider the following two functions, `fexample1` and `fexample2`.

```
function y = fexample1(x)
% Evaluates x^(1/3).
y = x^(1/3);
end
```

```
function y = fexample2(x)
% FEXAMPLE2(X)  Evaluates x^(1/3).
%               X can be a number, vector or matrix.
y = x.^(1/3);
end
```

If we use the `lookfor` command, it will not give use `fexample1` since the H1 line does not contain the function name, while both would be returned when we `lookfor evaluates`.

```
>> lookfor fexample
fexample2                    - (X)  Evaluates x^(1/3).

>> lookfor evaluates
fexample1                    - Evaluates x^(1/3).
```

```
fexample2                    - (X)  Evaluates x^(1/3).
evalmcw                      - Evaluates a list of functions in a editable text
     uicontrol.
```

HELP LINES are SUBSEQUENT COMMENT LINES without any breaks. The lines, in addition to the H1 lines, appear as the text when using `help` and should be a more detailed description of the function and its usage including what the input(s) and output(s) are. Using the MATLAB format, any reference to the FUNCTIONNAME and INPUT(S), OUTPUT(S) should be capitalized in the comments for formatting purposes though in reality they are not capitalized when used. Note that by formatting your help lines this way, in recent versions of MATLAB the capitalized function name appears in lowercase bold in the command window.

```
>> help fexample1
  Evaluates x^(1/3).

>> help fexample2
  `fexample2(X)`  Evaluates X^(1/3).
             X can be a number, vector or matrix.
```

To create paragraphs within help, have a line with only the comment symbol. An actually blank line will break the help comment block and the next comment is then not considered part of the help.

```
function A = fexample3(x,y)
% FEXAMPLE3(X,Y) evaluates X^Y for the input numbers or vectors X, Y.
%
% See also FEXAMPLE1, FEXAMPLE2.

% LAO
A = x.^y;
end
```

```
>> help fexample3
  fexample3(X,Y) evaluates X^Y
    for the inputs (numbers or matrices) X and Y.

  See also fexample1, fexample2.
```

What follows is the main block of the function that includes commands and comments. The function file can be with our without an `end`. The `end` is necessary for nested functions, or multiple functions in one file.

5.3. Function examples

This section discusses examples of functions; the best way to learn how to write functions is to use these functions as templates to do the exercises at the end of this chapter.

5.3.1 Basic function examples

The following examples were shown in Section 5.2 but are repeated here:

```
function y = fexample1(x)
% Evaluates x^(1/3).

y = x^(1/3);
end
```

The writing of `fexample1` could be improved. The HELP lines could include the function name, and it could be written so x could be a vector or matrix (if so desired). The improvement is seen in `fexample2` below.

```
function y = fexample2(x)
% FEXAMPLE2(X)  Evaluates X^(1/3).
%             X can be a number, vector or matrix.

y = x.^(1/3);
end
```

5.3.2 More function examples – multiple inputs

Multiple inputs of a function are listed with a comma separating the variable names. The function `fexample3` below shows how to create fancier HELP lines than `fexample1` and `fexample2`. Notice also that the function requires TWO inputs. When there are multiple inputs and/or outputs, it is especially important to make the HELP lines useful and clear for the order of the variables.

```
function A = fexample3(x,y)
% FEXAMPLE3(X,Y) evaluates X^Y
%   for the inputs (numbers or matrices) X and Y.
%
% See also FEXAMPLE1, FEXAMPLE2.

% LAO
A = x.^y;
end
```

The ability of having optional inputs is discussed in Section 6.8.

5.3.3 Multiple outputs

Multiple outputs can be handled in several ways, and one way may be better than the other depending on your usage of that function. See `quadratic1.m` and `quadratic2.m` for some examples. In one, you need to know there are multiple outputs and store them to see both of them. In the other, the multiple outputs are such that both are output.

```
function [r1,r2] = quadratic1(a, b, c)
% QUADRATIC1 calculates the quadratic formula on coefficients A, B, and C.
%                   [R1, R2] = QUADRATIC1(A, B, C) to compute
%                   the two roots, R1 and R2 of the quadratic equation
%                   Ax^2 + Bx + C = 0

% LAO, demonstrating multiple outputs
   d = b^2-4*a*c;

   r1 = (-b-sqrt(d))/(2*a);
   r2 = (-b+sqrt(d))/(2*a);
```

```
function [R] = quadratic2(a, b, c)
% QUADRATIC2 calculates the quadratic formula on coefficients A, B, and C.
%                   [R] = QUADRATIC2(A, B, C) to compute the two roots,
%                   R1 and R2 of the quadratic equation
%                   Ax^2 + Bx + C = 0.
%                   Returns a column vector R = [R1; R2]

% LAO, demonstrating multiple outputs
   d = b^2-4*a*c;

   r1 = (-b-sqrt(d))/(2*a);
   r2 = (-b+sqrt(d))/(2*a);

   R=[r1; r2];
```

Note that the above functions do not have `end`. As mentioned above, this is not necessary if there are not nested or multiple functions within the same file. One may argue that it is good practice to still include them. Also note that in `quadratic2`, the one output variable name is in brackets. This is not necessary; the functionality would be the same if the output variable name appeared without the brackets (as it does in `fexample1`.

The real difference in these functions is how the output variable(s) are defined and thus how we must use the functions correctly.

```
>> quadratic1(1,3,2)
ans =
    -2
```

```
>> quadratic2(1,3,2)
ans =
    -2
    -1
```

Note for quadratic1, only the first output variable is displayed and stored in ans, while for quadratic2, both solutions are displayed and stored in ans as a (column) vector. We could also store the output as a row vector as in quadratic2b.

```
function [R] = quadratic2b(a, b, c)
% QUADRATIC2B calculates the quadratic formula on coefficients A, B, and C.
%                    [R] = QUADRATIC2B(A, B, C) to compute the two roots,
%                    R1 and R2 of the quadratic equation
%                    Ax^2 + Bx + C = 0.
%                    Returns a row vector R = [R1, R2]

% LAO, demonstrating multiple outputs
   d = b^2-4*a*c;

   r1 = (-b-sqrt(d))/(2*a);
   r2 = (-b+sqrt(d))/(2*a);

   R=[r1, r2];
```

```
>> quadratic2b(1,3,2)
ans =
    -2    -1
```

The issue with the way these functions are written is that the user needs to know how the outputs are going to be displayed in order to use them correctly for anything useful. See the example below.

```
>> [x1,x2]=quadratic1(1,3,2)
x1 =
    -2
x2 =
    -1
>> [x1,x2]=quadratic2(1,3,2)
Error using quadratic2
Too many output arguments.
>> X=quadratic1(1,3,2)
X =
    -2
>> [X]=quadratic1(1,3,2)
X =
    -2
>> X=quadratic2(1,3,2)
```

```
X =
     -2
     -1
```

Examples of how to improve these functions are shown in the following chapter once conditional statements are discussed.

5.3.4 Bad examples

There are at least two things wrong with the following function. Can you find the errors?

```
function y = badfunction1(x)
% FCNEX1 Evaluates the cube root of a number X.

% LAO

x^(1/3)
```

Below we show a few commands with `badfunction1`.

```
>> help badfunction1
  FCNEX1 Evaluates the cube root of a number X.

>> badfunction1(8)
ans =
     2
```

Can you find anything wrong with `badfunction2`?

```
function Y = badfunction2(x)
% BADFUNCTION2  Evaluates the cube root of a number X.
% LAO

Y = x^(1/3)
end
```

Again, you may be able to figure out the errors by seeing the function in use.

```
>> help badfunction2
  badfunction2  Evaluates the cube root of a number X.
  LAO

>> badfunction2(8)
Y =
     2
ans =
     2
```

5.4. Exercises

1. Write a function named `ch5prob1` for the function $f(x) = \frac{x^5\sqrt{2x+3}}{(x^2+1)^3}$. The input of the function should be x and the output should be the calculated value $f(x)$. Write the function so that x can be a number, vector, or matrix just as the `sin` function works.
2. Consider the function

$$f(x) = \frac{1}{\sigma\sqrt{2\pi}} e^{-\frac{1}{2}\cdot\left(\frac{x-\mu}{\sigma}\right)^2}.$$

   (a) Create a function called `npdf` that takes as input x, μ, and σ. It then calculates $y = f(x)$, making sure that the calculations can be done if x is a number, vector or matrix, just as the `sin` function works (σ and μ are numbers and it is assumed the user will do this correctly; no error checking is done on this).
   (b) For $\mu = 0$ and $\sigma = 1$, graph $y = f(x)$ for $x \in [-5, 5]$.
   (c) Graph the following two plots on the same figure. The first should be of $y = f(x)$ for $x \in [130, 170]$, $\mu = 150.24$, and $\sigma = 8.44$. The second should be for $y = f(x)$ for $x \in [130, 170]$, $\mu = 153.07$, and $\sigma = 9.24$. Have a legend for this, with the first plot having the title "Verbal Reasoning" and the second plot having the title "Quantitative Reasoning". The overall title should be "GRE Test Scores July 1, 2015 to June 30, 2018" [5, p. 18].
3. Consider the function

$$f(t) = \frac{\Gamma\left(\frac{\nu+1}{2}\right)}{\sqrt{\nu\pi}\,\Gamma\left(\frac{\nu}{2}\right)}\left(1 + \frac{t^2}{\nu}\right)^{-\frac{\nu+1}{2}}$$

   where $\nu > 0$ and $\Gamma(x)$ is the Gamma function (`gamma(x)` in MATLAB)

$$\Gamma(x) = \int_0^\infty t^{x-1}e^{-t}\,dt.$$

   (a) Create a function called `tpdf` that takes as input t and ν and computes $y = f(t)$ making sure that the calculations can be done if t is a number, vector or matrix, just as the `sin` function works (no error checking is done on ν).
   (b) Perform the command `help tpdf`.
   (c) Use your function to graph $y = f(t)$ for $t \in [-4, 4]$ and $\nu = 1$.
   (d) Use your function to graph the following three plots on the same figure. They should all be for $t \in [-4, 4]$. The first plot should be with $\nu = 1$, the second with $\nu = 2$, and the third with $\nu = 5$. Have all be different colors and have a legend for the plots. The Greek letter ν can be written in text as "`\nu`" in MATLAB.

4. In Section 5.3.3 starting on page 87 there are four m-files: `quadratic1.m`, `quadratic2.m`, `quadratic2b.m` and `quadratic3.m`. All do *basically* the same thing but in slightly different ways. Save these functions to run them for the following problems.
   (a) Run `quadratic1` using the quadratic equations x^2+2x+1 and x^2-2x+5 by typing in just the function name with the appropriate inputs on the command line. What are your outputs, respectively?
   (b) Now run `quadratic1` on the same equations, but this time by typing in: `A = quadratic1`... and also `[r1, r2] = quadratic1`... How is the output different?
   (c) Now run do parts (a) and (b) but on `quadratic2` and `quadratic2b`. How does these differ from using`quadratic1`? Look at the files to see exactly how these differences occur. (Do not describe the differences that you see in the program code, but the differences in the output.)
   (d) Do parts (a) and (b) on `quadratic3`. Describe how this is different than `quadratic1`, `quadratic2`, and `quadratic2b`. (Again, do not describe the differences in the code, but the differences in the output.)
5. Using `quadratic3.m`, what can you say about the "correctness" of your output for the following equations? (Figure out with paper and pencil what the answers should be, and compare with the answers given by MATLAB.)
   (a) $x^2+\frac{x}{2}-\frac{15}{2}=0,$
   (b) $x^2+9.15x+15.05=0,$
   (c) $x^2-100{,}000{,}000x+1=0.$
6. Recall that polar coordinates (r,θ) can be converted to Cartesian coordinates (x,y) with the equations

$$x=r\cos\theta,$$
$$y=r\sin\theta.$$

   Consider the family of curves given by $r=1+c\,\sin(n\theta)$, where c is any real number and n is a positive integer. Write a function file called `pcurves` that takes as input n, c and θ where θ can be either a number or vector, n is a positive integer and c is any real number (no need to have error checks). The outputs should be the x and y values. Then, using your function, graph the curves (in the Cartesian (x,y) coordinate system) using `subplot` and varying n and c to answer the following questions. Be sure to label your graphs appropriately.
   (a) How do the graphs change as n increases?
   (b) How do the graphs change as c changes?
7. **Elementary row functions** For the following functions, make sure the help lines make it clear not only what the function does, but what the inputs are and in what order they should be.

(a) Create a function called `swap` that performs the first elementary row function on a matrix. The inputs should be `A,i,j` and the function should return a matrix `B` that is the same as `A` except the `i`th and `j`th rows are swapped.

(b) Create a function called `multrow` that performs the second elementary row function on a matrix. The inputs should be `A,r,i` and the function should return a matrix `B` that is the same as `A` except the `i`th row of `A` is multiplied by `r`.

(c) Create a function called `addrow` that performs the third elementary row function on a matrix. The inputs should be `A,r,i,j` and the function should return a matrix `B` that is the same as `A` except `r` times the `i`th row of `A` is added to the `j`th row.

8. You want to create multiple versions of homework and exam problems quickly. For example, you want to create homework problems with different coefficients of polynomials and you will use MATLAB to generate these coefficients. MATLAB has ways to create random numbers, but we want specific constraints and we may modify these constraints for different problems, so we will create functions to use repeatedly. The help lines for these functions should be very clear on how to use these functions. In Exercise 10 in Chapter 6 we will expand on these functions.

(a) Create a function `randInt2` that takes as inputs 2 integers, a and b. The function generates a random *integer* from a to b. No error checking will be done on the inputs a and b; we will assume the user inputs integers $a < b$. (Error checks will be created in Exercise 10 in Chapter 6.)

(b) Create a function `randList` that takes as input a vector of numbers (do not do any error checking on this; assume a vector is correctly input by the user). The function will return a random element from the vector (list).

CHAPTER 6

Control Flow

6.1. Relational and logical operators

Relational and logical operators allow one to compare two values, variables, etc. The operators take on the form `expression1 OPERATOR expression2` and evaluate to a logical data type; either `true` (1) or `false` (0).

Relational operators

You can use relational operators on vectors to compare to other vector(s) of the same size. The comparison is made component-wise and returns a vector of the same size with each entry either `true` or `false` (see Table 6.1).

```
>> 1 < 2
ans =
  logical
   1
>> 1 > 2
ans =
  logical
   0
>> 6/3 == 2
ans =
  logical
   1
>> 1 == sin(pi)
ans =
  logical
   0
>> A=randi([0,5],1,5), B=randi([-5,5],1,5)
A =
     4     1     3     4     5
```

Table 6.1 Relational operators.

<	less than
>	greater than
<=	less than or equal
>=	greater than or equal
==	equal
~=	not equal

Programming Mathematics Using MATLAB®
https://doi.org/10.1016/B978-0-12-817799-0.00011-9

```
B =
     5     1    -4    -4    -3
>> A <= B
ans =
  1×5 logical array
   1   1   0   0   0
>> A == B
ans =
  1×5 logical array
   0   1   0   0   0
>> A ~= B
ans =
  1×5 logical array
   1   0   1   1   1
```

Likewise, you can use relational operators to compare matrices of the same size.

```
>> X=randi([-5,5],2,3), Y = randi([-5,5],2,3)
X =
     1     5     3
     1    -2     3
Y =
    -1    -5     0
     1    -5     3
>> X > Y
ans =
  2×3 logical array
   1   1   1
   0   1   0
```

Logical operators

Logical operators can be used to create compound statements that evaluate to either `true` or `false` (see Table 6.2).

Eager versions will evaluate both expressions no matter what. **The short-circuit versions are only good on scalars and will only evaluate the second expression**

Table 6.2 Logical operators.

&, &&	AND (eager, short-circuit)
|, ||	OR (eager, short-circuit)
~	NOT
`xor`	exclusive OR
`all`	TRUE if all elements of array are TRUE
`any`	TRUE if any of elements of array are TRUE

if needed. Pros and cons to using each version are more apparent in complicated programs. For a discussion on the topic, search "short-circuit evaluation" on Wikipedia.

```
>> z=-exp(1), w = pi
z =
   -2.7183
w =
    3.1416
>> z > 0 && w > 0
ans =
  logical
   0
>> z > 0 || w > 0
ans =
  logical
   1
>> z > 0 & w > 0
ans =
  logical
   0
>> z > 0 | w > 0
ans =
  logical
   1
```

The eager versions of the logical operators produce matrices of 0s and 1s if one of the arguments is a matrix.

```
>> x = [-1 0 pi]; y = [1 2/3 -sqrt(2)];
>> x > 0 & y > 0
ans =
  1×3 logical array
   0   0   0
>> x > 0 && y > 0
Operands to the || and && operators must be convertible to logical scalar values.
>> X=randi([-5,5],2,3), Y=randi([-5,5],2,3)
X =
    -4     5     3
     5     0    -4
Y =
    -1     3     2
     5     5    -5
>> X > Y
ans =
  2×3 logical array
   0   1   1
   0   0   1
>> (X > Y) & (Y < 1)
```

```
ans =
  2×3 logical array
   0   0   0
   0   0   1
>> (X > Y) | (Y < 1)
ans =
  2×3 logical array
   1   1   1
   0   0   1
>> xor(X > 0, Y > 0)
ans =
  2×3 logical array
   0   0   0
   0   1   1
```

The any and `all` operators are useful, and work on vectors or on the columns of a matrix.

```
>> x=randi(10,1,5), y=randi(10,1,5)
x =
     5     5     4    10     4
y =
     2     8     4     3     5
>> any(y > x)
ans =
  logical
   1
>> all(y > x)
ans =
  logical
   0
>> all(X > 5)
ans =
  1×3 logical array
   0   0   0
>> all(Y >= 3)
ans =
  1×3 logical array
   0   1   0
>> any(Y < 0)
ans =
  1×3 logical array
   1   0   1
```

Note how any and `all` work column-wise or over all elements on matrices.

```
>> any(Y == 5, 2)
ans =
```

```
  2×1 logical array
   0
   1
>> any(X == 2, 'all')
ans =
  logical
   0
>> any(X == 0, 'all')
ans =
  logical
   1
```

Also note where the operators fall under the order of precedence (see Table 6.3).

6.2. If statements

Flow control in the form of `if` statements is vital for programming. MATLAB® has `if`, `if-else`, and `if-elseif-else` statements. These statements can be nested. Notice that there is no `begin` in these statements, but there must be an `end`.

If statements have the following pattern:

```
if SOME TEST EXPRESSION
        MATLAB command(s)
end
```

```
a=5; b=-2;
if b ~= 0 && (a == 5 || a == 3)
    yy = a/b;
    zz=a+b;
    disp('hi!')
end
```

Table 6.3 Precedence order of operators.

1	Parentheses
2	Exponents and transpose (')
3	Negation
4	Multiplication and division
5	Addition and subtraction
6	Colon operator
7	Relational operators
8	Logical AND (&, &&)
9	Logical OR (|, ||)
10	Assignment (=)

The following function `ispos` uses an `if-else` statement.

```
function y = ispos(n)
% ISPOS(N) returns true if N is positive, false otherwise.
    if n > 0
        y = true;
    else
        y = false;
    end
end
```

Example 6.2.1. Use an `if-else` statement to define a function for the following piecewise function:

$$f(x) = \begin{cases} 2x - 5 & x > 3, \\ x^2 - 8 & x \leq 3. \end{cases}$$

```
function f = pwexample(x)
% PWEXAMPLE(X) demonstrates an if-else statement for piecewise function.

    if x > 3
        f = 2*x - 5;
    else
        f = x^2 - 8;
end
```

NOTE: the above example will not work if x is a vector or matrix, even if component-wise calculations are used. See Example 6.4.1 on how to accommodate this.

Consider the following nested statement.

```
if x < 0
        s = -1;
else
        if x > 0
                s = 1;
        else
                s = 0;
        end
end
```

Nested statements such as the one above are common in other languages. In MATLAB as in some other languages, one can instead take advantage of `if-elseif-else` statements to accomplish the same thing but more efficiently.

```
if x < 0
   s = -1;
elseif x > 0
   s = 1;
else
   s = 0;
end
```

6.3. Switch/case

The `switch` command is an alternative to using nested statements or if-else-elseif statements in some cases. Switch tests against the equality of an expression against certain value(s). Switch statements work best against a discrete set of values, while in the case of the piecewise function example with inequalities, one would still need an if-statement.

```
z = input('Enter an integer: ');
switch mod(z,2)
    case 0
        disp('you entered an even integer')
    case 1
        disp('you entered an odd integer')
    otherwise
        disp('you did not enter an integer')
end
```

```
switch x
    case {-1, 0, 1}
        y = 3;
    case {-2, 2}
        y = 5;
    otherwise
        y = 7;
end
```

6.4. Use of characteristic functions

In mathematics, characteristic functions or indicator functions are functions of the form $\chi : A \rightarrow \{0,\ 1\}$ where A is a set within a larger set U. For any $x \in U$, $\chi(x) = 1$ if $x \in A$, and $\chi(x) = 0$ if $x \notin A$. For example, if $U = \mathbb{R}$ and $A = [0, \infty)$, then $\chi(2.5) = 1$ and $\chi(-3) = 0$.

Because the logical true is also 1, and logical false is 0 within MATLAB, we can easily create characteristic functions that can be implemented to create piecewise functions without if-statements, among other things. Consider the following example.

Example 6.4.1. Rewrite the piecewise function in Example 6.2.1 with the use of a characteristic function.

```
function f = pwexample2(x)
% PWEXAMPLE3(X) demonstrates an characteristic function to create a
% piecewise function.
%     This function will work if X is a vector or matrix.

    f = (2*x - 5).*(x > 3) + (x.^2 - 8).*(x <= 3);

end
```

6.5. For loops

For loops are extremely useful when you want to cycle through data and perform tasks, or when there is an iterative process. They have the following format:

```
for VARIABLE = EXPRESSION
     STATEMENT(S) OR COMMAND(S)
end
```

The expression is usually a vector defined with the colon operator or explicitly defined. It can also be a matrix, but note that implementation is done one column at a time.

Consider the following function, `myscatter`, which creates n random points (x, y) and plots each one.

```
function myscatter(n)
% MYSCATTER Example of using a for loop to create a scatter plot.
%    MYSCATTER(N) creates a plot of N points that are randomly generated.
    x=rand(1);
    y=rand(1);
    plot(x,y, '*', 'MarkerSize',8)
    hold on
    for k=2:n
        x=rand(1);
        y=rand(1);
        plot(x,y, '*', 'MarkerSize',8)
    end
    hold off
end
```

Note that this is not the most efficient way of doing this within MATLAB. First, there is already a `scatter` plot command and secondly, we can make the above code more efficient by what is called VECTORIZING THE CODE, which we will discuss later.

Another common use of a for loop is for calculating sums.

```
function s = geomseries(r, n)
% GEOMSERIES computes the sum of the geometric sequence
%            Y = GEOMSERIES(R,N) is the sum of R^k from k=0 to k=N.
    s = 0;
    for k=0:n
        s = s + r^k;
    end
end
```

The following example shows how to use nested for-loops to work with matrices.

Example 6.5.1. Improve the previous function in Example 6.2.1 so that it also works if x is a vector or matrix with the use of if-statements and nested for-loops.

$$f(x) = \begin{cases} 2x - 5 & x > 3, \\ x^2 - 8 & x \leq 3. \end{cases}$$

```
function f = pwexample3(x)

f = 0*x;
[rows,cols] = size(x);
for j = 1:rows
        for k = 1:cols
                if x(j,k) > 3
                    f(j,k) = 2*x(j,k) - 5;
                else
                    f(j,k) = x(j,k)^2 - 8;
                end
        end
end
```

We saw in Example 6.4.1 that this longer code is not necessary with the use of characteristic functions, but there are many examples why nested for loops would be necessary. For example, if the index of the matrix entry is needed, or in the case of double (triple, etc.) summations.

The colon operator can be used to have the for loop increment backwards, or you can explicitly state the values as seen in the examples below.

```
for k = 10:-1:1
    disp(k)
    pause(1)
end
disp('Blast off!')
%% different increment
for k = 2:2:8
    disp(k)
    pause(0.5)
end
disp('Who do we appreciate?')
%% using defined vector
x = [2 4 6 8];
for k = x
    disp(k)
    pause(0.5)
end
disp('Who do we appreciate?')
```

The following example shows how the for loop works with matrices. Run the code to see the results. Note that the results within the loop are displayed for demonstration purposes.

```
x=[2 6; 4 8];
for k=x
   disp(k)
   pause(0.5)
end
disp('Who do we appreciate?')
%% showing the matrix implementation
x
for k=x
    k
    sumk = sum(k)
    maxk = max(k)
end
```

6.6. While loops

While loops can be used to accomplish similar and different tasks as a for loop. The format of a while loop is as follows.

```
while EXPRESSION
        STATEMENT(S) OR COMMAND(S)
end
```

The statements or commands within the loop are executed as long as the expression is true. This can mean that until something with expression changes, the loop may become an infinite loop.

You can use while loops to generate better data, or continue to ask the user until a valid answer is submitted.

```
a = input('Enter a non-zero integer: ');
while (a == 0) | (mod(a,1) ~= 0)
        a = input('Enter a non-zero integer: ');
end
b = randi([-10,10]);
c = randi([-3,3]);
while (b^2 - 4*a*c <= 0)
        b = randi([-10,10]);
        c = randi([-3,3]);
end
```

While loops are also useful in working with tolerance checks. In general, the difference between while loops and for loops is with for loops, the number of times to loop through is known, and with while loops, you want to keep doing something until something changes.

6.7. Useful commands break, continue, return, and error

The `break` command terminates the loop and the first statement after the loop's `end` is executed. If `break` is used within nested loops, the inner loop is executed and control is passed to the loop at the next higher level.

The `continue` command allows the call of the command between the `continue` statement and the `end` for that loop to be bypassed, and the next increment of the loop is then executed. Note the differences when the following code is run.

```
%% break example
clc
for k=1:5
        if k==3
                break
        disp('ah ha!')
        end
        disp(k)
end
%% continue example
clc
for k=1:5
        if k==3
                continue
```

```
        disp('ah ha!')
        end
        disp(k)
end
```

The `return` command is useful for script or function files, in which the command acts like an `end` to the function, or "end of file" of the script file.

The `error` command is useful in which you can develop meaningful error messages to the user. One can accomplish a similar task using `disp` or even `fprintf` (more on that command later) along with `return`, but `error` is more efficient and the error message is automatically displayed in red similar to MATLAB errors.

```
n = input('Enter a non-zero number: ')
if n == 0
        error('You entered zero. Try again.')
end
```

6.8. Optional inputs and outputs of functions

One can work with optional inputs and outputs using `nargin`, `varargin`, `nargout`, and `varargout` variables. The `nargin` variable is a number that gives you how many inputs were given to the function, while `nargout` is the number of output arguments requested by the user. These variables allow us to use functions in multiple ways. The use of the `nargout` variable is best explained with the SIZE command in MATLAB:

```
>> A
A =
     1     2     3
     4     5     6
>> size(A) % no output specified
ans =
     2     3
>> y = size(A) % one output specified
y =
     2     3
>> [r, c] = size(A) % two outputs specified
r =
     2
c =
     3
```

The following function `quadratic4` was written to display or store the outputs similar to the `size` command but for the quadratic function.

```
function [r1,r2] = quadratic4(a, b, c)
% QUADRATIC4 uses the quadratic formula.
%                   [R1, R2] = QUADRATIC4(A, B, C) to compute
%                   the two roots, R1 and R2 of the quadratic equation
%                   Ax^2 + Bx + C = 0.
    d = b^2-4*a*c;

    r1 = (-b-sqrt(d))/(2*a);
    r2 = (-b+sqrt(d))/(2*a);

    if nargout < 2
        r1 = [r1, r2];
    end
end
```

The function `quadratic5` demonstrates the use of `nargin` in addition to `nargout`. This function allows for either two or three input variables for the coefficients of a quadratic equation. If two coefficients are input, the leading coefficient a is set equal to 1.

```
function [r1,r2] = quadratic5(n1,n2,n3)
% QUADRATIC5(A,B,C) or QUADRATIC5(B,C) uses the quadratic formula for the given
     coefficients.
%                   [R1, R2] = QUADRATIC5(A, B, C) to compute
%                   the two roots, R1 and R2 of the quadratic equation
%                   Ax^2 + Bx + C = 0.
%                   [R1, R2] = QUADRATIC5(B, C) then A = 1.
    if nargin == 2
        a = 1;
        b = n1;
        c = n2;
    else
        a = n1;
        b = n2;
        c = n3;
    end

    d = b^2-4*a*c;

    r1 = (-b-sqrt(d))/(2*a);
    r2 = (-b+sqrt(d))/(2*a);

    if nargout < 2
        r1 = [r1, r2];
    end
end
```

The variables varargin and varargout are cell arrays of variable input or output arguments. The quadratic6 function demonstrates the use of varargin to function similarly to quadratic5.

```
function [r1,r2] = quadratic6(varargin)
% QUADRATIC6(A,B,C) or QUADRATIC6(B,C) uses the quadratic formula for the given
    coefficients.
%                   [R1, R2] = QUADRATIC6(A, B, C) to compute
%                   the two roots, R1 and R2 of the quadratic equation
%                   Ax^2 + Bx + C = 0.
%                   [R1, R2] = QUADRATIC6(B, C) then A = 1.
    if nargin == 2
        a = 1;
        b = varargin{1};
        c = varargin{2};
    else
        a = varargin{1};
        b = varargin{2};
        c = varargin{3};
    end

    d = b^2-4*a*c;

    r1 = (-b-sqrt(d))/(2*a);
    r2 = (-b+sqrt(d))/(2*a);

    if nargout < 2
        r1 = [r1, r2];
    end
end
```

The function mySquare demonstrates another nice use for varargin. This function allows for optional LineSpec (line style, marker symbol, and color).

```
function mySquare(s,x,y,varargin)
% MYSQUARE plots a square with length S with lower left corner at X,Y
%     MYSQUARE(S,X,Y) or MYSQUARE(S,X,Y, LINESPEC)

plot([x,x+s,x+s,x,x],[y,y,y+s,y+s,y],varargin{:})
```

The function myRandN is another demonstration of using both nargin and varargin.

```
function y=myRandN(mu,sigma,varargin)
% MYEXAMPLE demonstrates using nargin and vargin
switch nargin
    case 2
```

```
        y=randn*sigma + mu;
    case 3
        y=sigma*randn(varargin{1}) + mu;
    case 4
        y = sigma*randn(varargin{1},varargin{2}) + mu;
end
end
```

6.9. Exercises

1. Write a function named `hw6prob1` for the function $f(x) = \dfrac{x^3\sqrt{2x+3}}{(x^4+1)^3}$ that does the following:
   - The input of the function should be x and the output should be the computed value $y=f(x)$.
   - Write the function so that x can be a number, vector or matrix.
   - Using the `any` and/or `all` logical operators, first test within the function to make sure that all of the values of x are valid for $\mathbb{R}$-valued calculations. If not, an appropriate error message is displayed using the `error` command.
   - You should test your calculations with at least 4 different inputs. Write down the input tests you used, along with the answers that your function gives you.
2. Create a function called `isZ` that will take one input and output a **logical** true or false (value of either 0 or 1 specifying false or true, respectively). If the input is an integer (NOT TALKING ABOUT DATA TYPE INTEGER – meaning it is not π, or not 3/2, etc.), then it will return a logical `true`, otherwise it will return a logical `false`. Note that this should still work if the input is a vector or matrix, and would return a vector or matrix of true/false based on whether each entry/component is an integer or not. Run tests of the function. This function should be used on any subsequent problem that needs to check if a variable or input is an integer.
3. Create a function called `isEven` that will have one input that is supposed to be an integer. First check whether the input is an integer; if not, use the `error` command to display an appropriate error message. The function should return a logical true if the input is even, and a logical false if it is odd. Run tests of the function.
4. Finish the function `plotVec` that is found on the text website. This function has one or more inputs and uses `varargin`. The first input is `P`, which is a *column* vector with two or three rows (check for it: if not, return an error). The function will take the column vector and plot the vector as a line segment from the origin to the given point (from the input `P`). Any other input arguments are optional and specify how the vector will be drawn (color, line width, marker type, etc.). The

function will not have an output (it will only create a plot). *Hint: remember the difference between* `plot` *and* `plot3`! Run it once for a 2D point and another time for a 3D opint, using your choice of plot specifications (or none) on each.
BONUS: create a function `plotVec2` that will expand on `plotVec`. This function will take a MATRIX of 2D or 3D points and connect them all, in order. Thus it could create a polygon from the matrix in which each column is a vertex of the polygon. Discuss the limitations or difficulties this may have in coding it and/or implementation.

5. Finish the function `plotPlane` that is found on the text website. This function has two or more inputs and uses `varagin`. The first input is `M` which is a matrix that has exactly three rows and at least two columns (check for it; if not, return an error). The second input `ax` determines the domain for the plane; the domain will be from `-ax` and `ax` (*Hint: remember* `meshgrid`!) Any other optional inputs will specify `EdgeColor`, etc. to plot the plane using the `mesh` command. Use the function to plot a plane using M =`randi`; create M, display it and then use your function to plot the plane.

6. Create a new function called `parity` that does the following:

$$f(x) = \begin{cases} 1 & x \text{ is even,} \\ 0 & x \text{ is odd,} \\ -1 & x \text{ is not an integer.} \end{cases}$$

We will be doing our own checking if x is an integer even if MATLAB has a function for this. Make sure x can be a number, vector, or matrix, and that the function then returns a corresponding number, vector, or matrix.

(a) Calculate $f(x)$ for $x = [2, -3, -4, 5, 0, 1/2]$.

(b) Calculate $f(x)$ for

$$x = \begin{bmatrix} 0 & -1 \\ \pi & 100 \end{bmatrix}.$$

7. Create a new function called `npdf2` that does the same as `npdf` in Problem 2 in Chapter 5, but IN ADDITION:

- The function npdf2 has OPTIONAL inputs σ and μ. If only one input is given (i.e., x is the only input), then $\sigma = 1$ and $\mu = 0$. If two inputs are given, they are x and σ and μ is taken to be equal to 0. If three inputs are given, they are x, σ, and μ. If the optional inputs are given, the following "error checks" are done.
- Checks to make sure the inputs σ, and μ are numbers rather than vectors or matrices. Here we will create our own check (even though MATLAB may have functions that can do these checks). If a vector or matrix was input, an appropriate error message is displayed.

- Checks to make sure $\sigma > 0$; if not, an appropriate error message is displayed.

8. Create a SCRIPT FILE called `pball.m` that will use the `input` command to take ask for the user, "How many Power Ball tickets?" (n). Check that n is a number (rather than a vector or matrix) and a positive integer and if not, keep asking the user "Try again. How many Power Ball tickets?" until the input is correct. Then using `input` again, ask, "Power Play option? (y/n)" (*PP*). Likewise, if the answer to the "Power Play" option is not "Y", "y", "N", or "n" keep asking the user, "Try again. Power Play option? (y/n)" You will use the `switch/case` command for the answer to the "Power Play" question. If *PP* is "Y" or "y", each Powerball ticket costs $3. If *PP* is "N" or "n", each Powerball ticket costs $2. The file then displays a matrix with n rows in which each row is a Powerball ticket (as specified in Exercise 9(d) of Chapter 2 [The first five entries of each ticket are the "white balls" and the last entry of each ticket is the "Powerball" for the drawing. For each ticket, the first five numbers should be random integers from 1 to 69 **with no repeats** and not necessarily in order and the last number should be a random integer from 1 to 26.] but can now be done in an easier way). The amount owed for the Powerball tickets is also displayed. You do not need to get fancy with displaying the tickets and price; we will work on that later.
Generate `randi(10)` Power Ball tickets and your choice as to whether or not the Power Play option is selected.

9. A circle of radius r and center (h, k) can be parameterized with the equations

$$x = r\cos(\theta) + h$$
$$y = r\sin(\theta) + k$$

for $\theta \in [0, 2\pi]$. Create a function `myCircle` that takes as inputs a positive number r, a 2D point (vector) C, and an optional positive integer n. Error checking should be done on r, C (and n if input) and appropriate messages displayed using the `error` command. This function will output vectors x and y of size n (with default value of $n = 100$) that are the calculated parametric equations for a circle with radius r with center $C(h, k)$. Have the output be similar to `size` or `quadratic4` (page 105) in that both are output regardless of how the user runs the function.

(a) Use your function to plot the unit circle and another circle with radius 3 and center $(-2, 4)$.

(b) Create another graph where you plot the unit circle but with no input for n, $n = 4$, $n = 5$, and $n = 9$.

The command `axis equal` may come in handy for these plots!

10. We will expand on the Problem 8 in Chapter 5.

(a) Create a function `randInt.m` that takes as inputs two OR three arguments. If the two arguments a and b are input, then the function generates a random integer from a to b. If the three arguments a, b, and c are input, the function generates a random number from a to b with an increment of c. So if I input `randInt(-5,5)` it would give me a random integer (NOT data type integer!) from -5 to 5. But if I input `randInt(-5,5,.5)` it would give me a random number from the list $-5, -4.5, -4, -3.5, \ldots, 4, 4.5, 5$. ERROR CHECK: check to make sure $a < b$ and (if input) $c > 0$. If either of these errors occur, use the `error` command to stop the function and output an appropriate error message.

(b) Create a function `nonzeroRand.m` that takes as inputs two OR three arguments. If the two arguments a and b are input, then the function generates a random NON-ZERO INTEGER from a to b. If the three arguments a, b, and c are input, the function generates a random NON-ZERO NUMBER from a to b with an increment of c. You will use a WHILE loop within this function. ERROR CHECK: check to make sure $a < b$ and (if input) $c > 0$. If either of these occur, use the `error` command to stop the function and output an appropriate error message.

11. You are creating a multiple choice problem in which students will be asked to solve $ax + b = cx + d$ for x, where a, b, c, and d are certain random integers. You have decided that the wrong answers given in the problem are:

$$w1 = \frac{b-d}{a-c}, \quad w2 = \frac{a-c}{d-b}, \quad w3 = \frac{b+d}{a-c}, \quad \text{and } w4 = \frac{b+d}{a+c}.$$

Initially, the problem is set such that a is a random integer between 2 and 6, b is a **non-zero** random integer between -4 and 4, c is a random integer between 3 and 7, and $d = a + b - c - 1$. The problem is set that if $c = a$ or $a = c + 1$, a new c is generated until both $c \neq a$ and $a \neq c + 1$. The problem is also set that if $b = d$, a new b is generated (and thus d) until $b \neq d$. (Think: why would we want this check? Unfortunately, you notice that there are times that some coefficient combinations make it so there are duplicate answers listed in the multiple choice problem. You already realize there are issues when the denominators and/or numerators equal zero but those are already taken care of; you need to find the other bad combinations. Instead of trying to figure out algebraically how to define the coefficients so this does not happen, you write a MATLAB script called `algProblem.m` to help pinpoint what combination(s) of coefficients will and will not work.)

(a) What is the correct answer to the problem in terms of a, b, c, and d? Show all work on paper.

(b) Create a script file `algProblem` that does the following:

- Initialize vectors A, B, and C to contain all of the possibilities of values for a, b, and c, respectively.

- Initialize a counter variable (name of your choice under MATLAB guidelines) to equal 0. This variable will keep track of how many "bad" coefficient combinations there are.
- Create nested `for` loops, the first based on the possibilities for a (the elements in A), the second based on the possibilities for c, and the third based on the possibilities for b (this is purposely "out of order"). These `for` loops will cycle through all of the possible combination of coefficients and perform the checks.
- Within the second `for` loop, check to see if any of the denominators and numerators of the answers that are based on a and c equal zero. If so, the `continue` command is used to move onto the next iteration of this `for` loop (and thus the next combo possibility for a and c). This combination is not "counted" since other code within the problem already disallows this combination.
- Within the third `for` loop define d based on the current values of a, b, and c and the formula given above.
  - A check is made to see if any of the denominators and numerators of the answers (both correct and wrong!) that are based on b and d equal zero. If so, again the `continue` command is used to move onto the next iteration of the loop (and thus the next set of coefficients) since similar to above this combination is already disallowed and not "counted" as a bad combination.
  - Define variables for the five answers to the multiple choice question (the correct and four wrong answers).
  - We want to make sure all five answers are distinct values. How many pairwise comparisons need to be done? There are multiple ways to do this check. One way is to create a logical vector that compares each pair of answers with each other. Thus this vector should contain only ones and zeros. For example, the first entry of the vector equals `true` (1) if the $w1 = w2$ and `false` (0) otherwise.
  - Using the `any` command on the vector, if any of the values in this logical vector equals `true` (1) (i.e., any pair of answers is equal), then the current combination of coefficients is "bad." The counter from above is incremented, the coefficients are output to the command window along with the five calculated answers, in the order of: correct, $w1$, $w2$, $w3$, and $w4$. Thus when the script is run, you can see by the end how many "bad" combinations there are, and you can scroll through and see not only the coefficients but the five answers of the multiple choice problem. Try to make the output as clear and concise as possible; you may want to create a vector `abcd` that has

all four values of the coefficients and output that vector; also create a vector `Answers` that has the multiple choice answers and output that vector.

(c) How many bad combinations of coefficients are there? What are they? State your answer on your paper and list the bad combinations of coefficients clearly. No need to show your work on this one; I should be able to run your script to see the work. Do not run the script file within the homework file.

12. Create a new function called `mysgn` that is a *variation* on the signum function:

$$\operatorname{sgn}(x) = \begin{cases} 1 & x > 0, \\ 0 & x = 0, \\ -1 & x < 0. \end{cases}$$

In this case, `mysgn` will have the following criteria:

- Has one input x which can be a number, vector, or matrix, and will have as output y that is a (1×4) row vector y.
- $y(1,1) = \begin{cases} 1 & x \text{ or all elements are } > 0, \\ -1 & x \text{ or all elements are } < 0, \\ 0 & \text{otherwise,} \end{cases}$
- $y(1,2) = \begin{cases} 1 & x \text{ or any elements are } > 0, \\ 0 & \text{otherwise,} \end{cases}$
- $y(1,3) = \begin{cases} -1 & x \text{ or any elements are } < 0, \\ 0 & \text{otherwise,} \end{cases}$
- $y(1,4) = \begin{cases} 0 & x \text{ or any elements are } = 0, \\ 1 & \text{otherwise.} \end{cases}$
- This function should run like the `size` function in MATLAB; if all four outputs are specified, it assigns $y(1,1)$ to the first specified output, and $y(1,2)$ to the second specified output, and if one or no outputs are specified, it gives a vector with containing the answers. Any other number of outputs specified returns an error with a meaningful message.

(a) Calculate `u = mysgn(x), v = mysgn(a), w = mysgn(b)` for $x = -\sqrt{2}$, $a = \pi$, and $b = 0$.

(b) Calculate `[a,b,c,d]= mysgn(x)` for `x = -10:2:10.`

(c) Calculate `mysgn(x)` for `x = 0:2:10.`

(d) Calculate `mysgn(x)` for `x = 1:2:10.`

(e) Calculate `mysgn` for `A = eye(3).`

(f) Calculate `mysgn` for `B = ones(3).`

(g) Calculate `mysgn` for `-3*A.*B`.

13. Notice what happens when we enter `(-8)^(1/3)` into MATLAB. Create a function `rootn` that gives us the real-valued third root of a negative integer. The function `rootn` will calculate $\sqrt[n]{x}$ and also does the following for a number x:

- The inputs are n and x of which we are going to compute $\sqrt[n]{x}$. The help lines should indicate what the inputs are and in what order.
- The function checks that x is a number rather than a vector or matrix. If not, an error message is displayed and the function should stop.
- Another check is that n is a natural number. If not, an appropriate error message should be displayed and the function should stop.
- If n is even, it checks to make sure that x is nonnegative. If not, an error message is displayed and the function should stop.
- If n is odd, it computes the root as expected. In other words, it should give -2 as the answer for $\sqrt[3]{-8}$.

NOTE that we are creating our own version of the MATLAB function `nthroot`.

CHAPTER 7

Miscellaneous Commands and Code Improvement

7.1. Miscellaneous commands

7.1.1 The fprintf command

The command `fprintf` allows us to specify more formatting, including new lines and tabs to display text and/or variables. The command `disp` can be used to do this, but it does not allow new lines, tabs, or special formatting.

```
>> n=5; x=1/2;
>> y=geomseries(x,n);
>> fprintf('The estimate of the geometric series when r = %f and n = %i is %f',x,n,y)
The estimate of the geometric series when r = 0.500000 and n = 5 is 1.968750>>
```

Notice if you copy the above `fprintf` command and run it, the command prompt and cursor is at the end of the text, which is difficult to see. To get a new line, use "\n". A tab is "\t".

```
>> fprintf('\n\nThe estimate of the geometric series when r = %f and n = %i\n\t is %f
    \n\n',x,n,y)


The estimate of the geometric series when r = 0.500000 and n = 5
        is 1.968750
```

Notice the "placeholders" for where the variables go in the text. These are using formats.

```
>> fprintf('\n Here is 2^10: %f\n\n',2^10)

 Here is 2^10: 1024.000000

>> fprintf('\n Here is 2^10: %e\n\n',2^10)

 Here is 2^10: 1.024000e+03

>> fprintf('\n Here is 2^10: %E\n\n',2^10)

 Here is 2^10: 1.024000E+03
```

Programming Mathematics Using MATLAB®
https://doi.org/10.1016/B978-0-12-817799-0.00012-0

What if you wanted to display only a certain number of decimal places?

```
>> fprintf('\n Here is e: %4.2f\n\n',exp(1))

 Here is e: 2.72

>> fprintf('\n Here is e: %2.4f\n\n',exp(1))

 Here is e: 2.7183
```

Further explanations of the above formatting can be found in the MATLAB® documentation.

The formatting in the `fprintf` does not pay attention to the formatting you have set outside of the command (see the section below for a table of formats):

```
>> format bank
>> exp(1)

ans =

          2.72

>> fprintf('\n Here is e: %f\n\n',exp(1))

 Here is e: 2.718282
```

You can get fancy with justification:

```
>> fprintf('\nsome text:%3.0f\nsome text:%3.0f\n\n',7,2^10)

some text:  7
some text:1024

>> fprintf('%-2.0f\n%-2.0f\n\n',7,2^10)
7
1024

>> fprintf('\nsome text:%-6.0f\nsome text:%-6.0f\n\n',7,2^10)

some text:7
some text:1024
```

Another nice format is using "%g" which will use either %f or %e, whichever is the shortest:

```
>> fprintf('%f %f\n\n', 2^10, 2^100)
1024.000000 1267650600228229401496703205376.000000
```

```
>> fprintf('%e %e\n\n', 2^10, 2^100)
1.024000e+03 1.267651e+30

>> fprintf('%g %g\n\n', 2^10, 2^100)
1024 1.26765e+30
```

7.1.2 The sprintf command

The command `sprintf` is the same as `fprintf` except that its output is a string, which can be useful for labeling plots (see Fig. 7.1).

```
a=10;
b=34/(2*pi);
plottitle=sprintf('Helix with a=%i and b approximately %3.5f',a,b);
t=linspace(0,100,750);
x=10*cos(t);
y=10*sin(t);
z=b*t;
plot3(x,y,z)
title(plottitle)
```

It is also useful for error messages.

```
>> n=101;
>> errormsg = sprintf('We need n to be even. You entered n = %i',n);
>> if mod(n,2)~=0
    error(errormsg)
end
We need n to be even. You entered n = 101
```

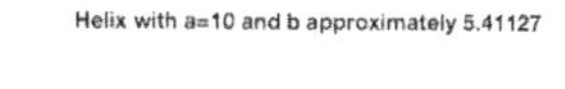

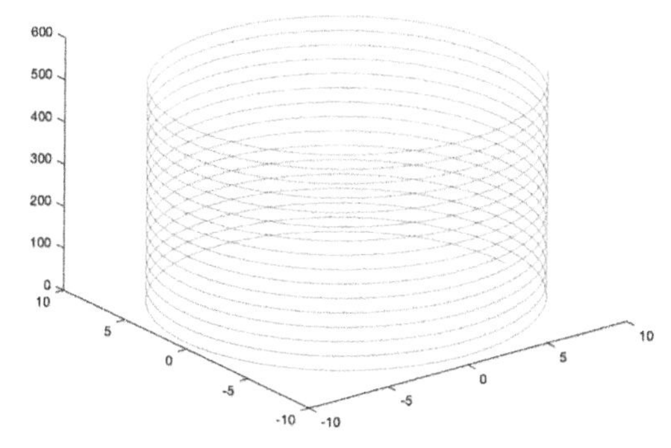

Figure 7.1 Example using `sprintf` with `title`.

7.1.3 Formats revisited

In Section 1.2.3 using `format` was discussed. Recall that the default format is to display four decimal places (which is actually `format short`). Above we had switched to `format bank`. To get back to the default format, use just `format`.

```
>> x=10*pi

x =

        31.42

>> format
>> x

x =

   31.4159
```

See Table 1.4 for other formats one can use.
Notice that `long e`, `short g`, etc. can be one word:

```
>> format compact
>> format longe
>> x
x =
     3.141592653589793e+01
>> format +
>> x
x =
+
>> -x
ans =
-
>> format shorteng
>> 1000*x
ans =
    31.4159e+003
>> pi/1575
ans =
     1.9947e-003
>> pi/22222
ans =
   141.3731e-006
```

7.1.4 The save/load commands

One can save workspace by using the command `save`, such as

```
>> save('saveddata1.mat')
```

Then one can load that file at any other time to retrieve the data.

```
>> load('saveddata1.mat')
```

7.1.5 The tic/toc commands

The command tic starts a timer, while toc ends it and stores or displays the amount of time used.

```
>> tic
>> n=0
n =
     0.0000e+000
>> for k=1:10000
n=n+n^2;
end
>> toc
Elapsed time is 12.720630 seconds.
```

Notice the above is also the time it took to type those commands into the command window.

7.1.6 The fill command

The fill(x,y,color) command fills the area between the points (x, y) given with the given color with a black edge, connecting automatically the first point with the last point. Fig. 7.2(A) below shows using fill on a sine wave and Fig. 7.2(B) fills a unit circle.

```
% First example
color1 = [0,0.4470,0.7410]; % default first color starting in version 2014a
x=linspace(-3,3);
fill(x,sin(pi*x),color1)
% Second example
t=linspace(0,2*pi);
xc = cos(t); yc = sin(t);
fill(xc,yc,color1)
```

If x and y are defined with x=linspace(-2,2); y=x.^2; you get Fig. 7.3(A) using fill(x,y,color1). For regions such as "area under the curve", you will need to adjust the points. Using fill([-2,x,2],[0,y,0],color1) on the same x and y leads to Fig. 7.3(B).

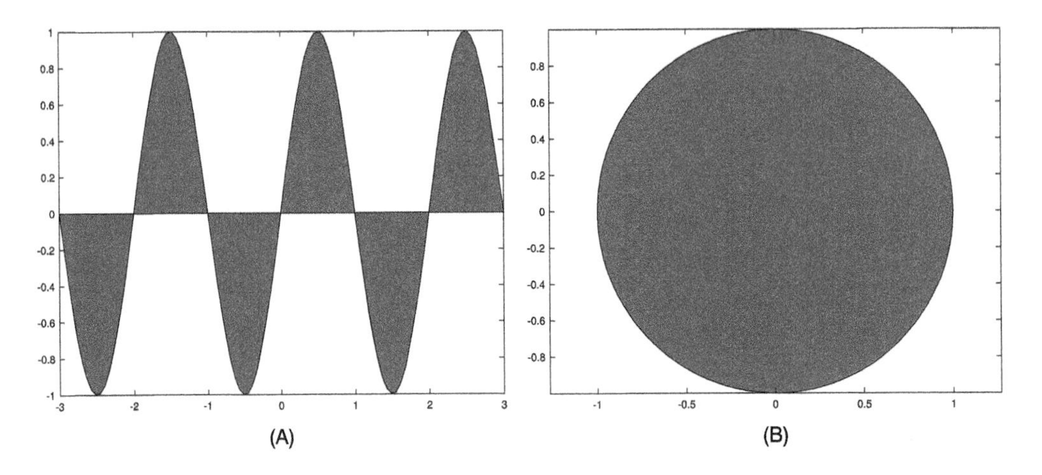

Figure 7.2 Examples of `fill` command. (A) Filling $y = \sin(\pi x)$, (B) Filling a circle.

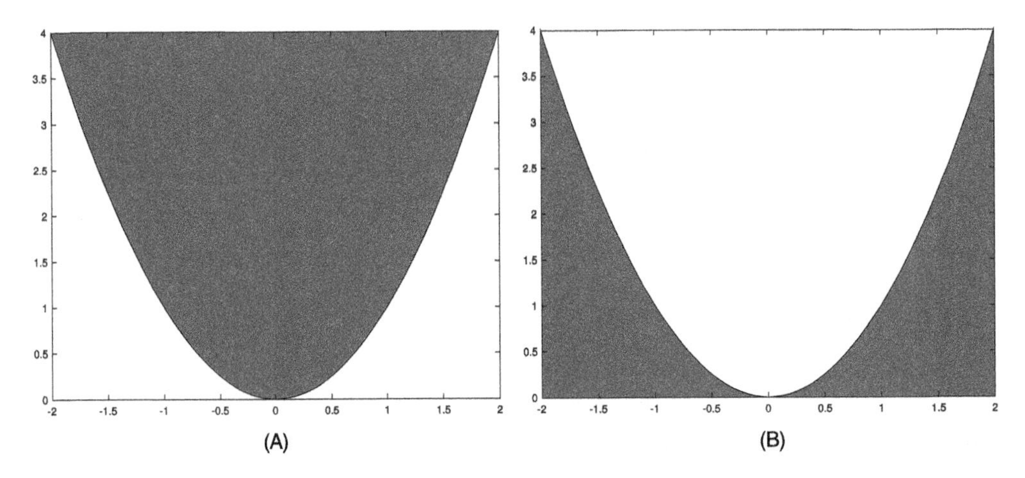

Figure 7.3 Adjusting to get the area under the curve. (A) Fill $y = x^2$, (B) Adding points to adjust.

Here follows another example of "area under the curve" that created Fig. 7.4.

```
color2=[0.8500, 0.3250, 0.0980]; % default 2nd color since v2014a
x=linspace(-pi,pi);
y=cos(x)+2;
fill([x(1), x,x(length(x))],[0, y,0],color2)
axis equal
```

For basic regions between two functions, one can do the following to create Fig. 7.5. Notice the use of the useful command `fliplr(x)` which creates a vector that has the values of x but in reverse order.

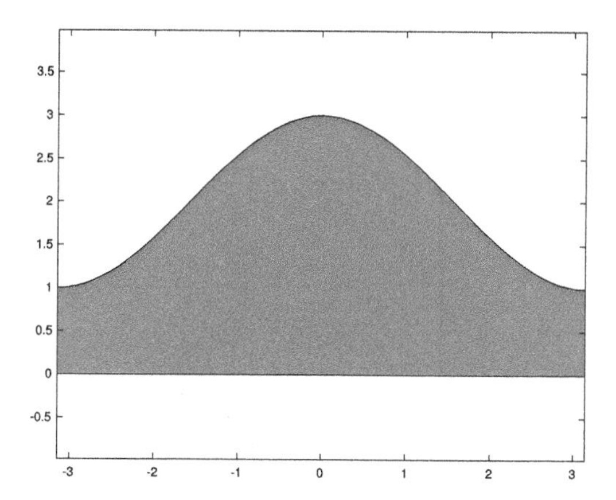

Figure 7.4 Another area under curve.

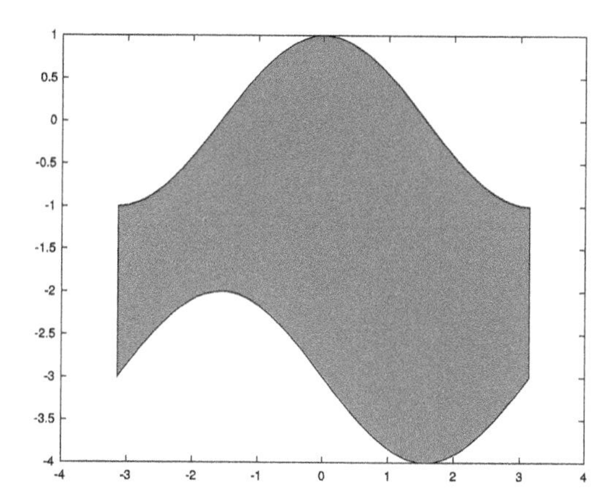

Figure 7.5 Basic region between two curves.

```
x=linspace(-pi,pi);
y=cos(x);
x2=fliplr(x);
y2=-sin(x2)-3;
fill([x,x2],[y,y2],color2)
```

Here is another example using `fill` (see Fig 7.6).

```
t = pi/8: pi/4: pi/8 + 2*pi;
xo=cos(t); yo=sin(t);
fill(xo,yo,'r')
hold on
plot(xo,yo,'w','LineWidth',10)
plot(xo,yo,'k')
```

```
hold off
axis equal
ax=1.2;
axis([-ax,ax,-ax,ax])
```

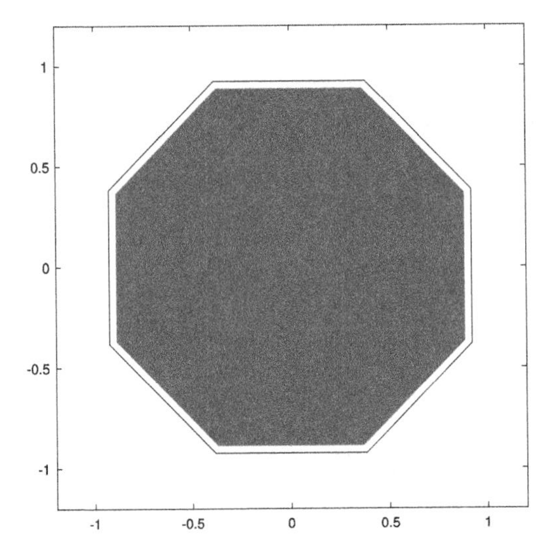

Figure 7.6 Using `fill` for stop sign example.

7.1.7 The command alpha

The order of plot commands, including `fill`, can make a difference on the appearance in the figure window. If you run the following code within MATLAB, since the `fill` comes after plotting the sine wave, it "covers up" the sine wave as shown in Fig. 7.7(A).

```
plot(xsine,ysine)
hold on
fill(xtriangle,ytriangle,'y')
hold off
```

If we reorder the commands as in the following code, the sine wave is then plotted "on top" of the filled triangle, as shown in Fig. 7.7(B).

For certain plotting commands, the `alpha(n)` command can create transparency. The value n is a number between 0 and 1, with 1 being opaque, and 0 being completely transparent. (Experiment!) Thus a combination of the order of the commands and the use of alpha can give you more control over the look of the figure (see Fig. 7.8).

```
plot(xsine,ysine,'k')
hold on
```

```
fill(xtriangle,ytriangle,'g')
alpha(0.3)
hold off
axis equal
```

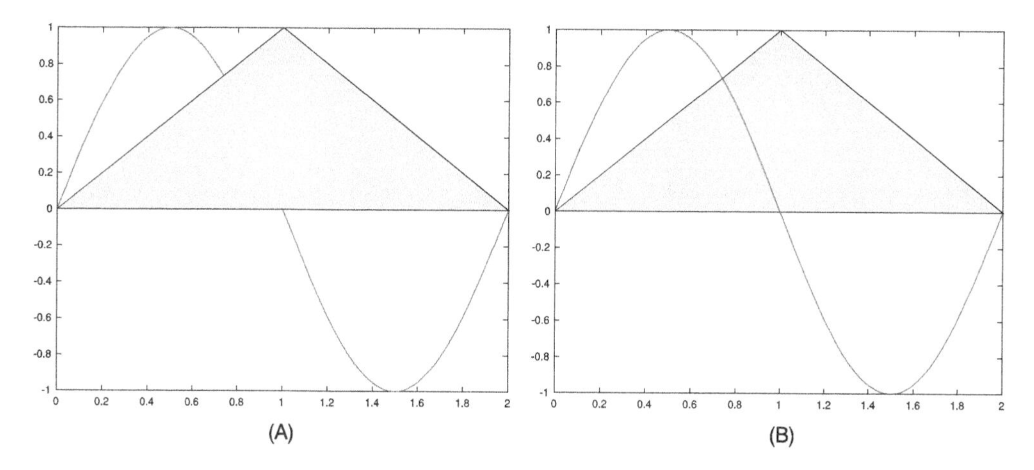

Figure 7.7 The order of plot and fill commands matter. (A) Plot first, then fill, (B) Fill first, then plot.

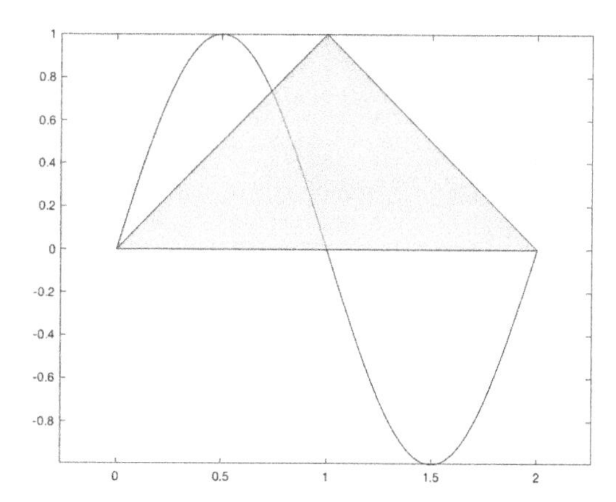

Figure 7.8 Using `alpha` to create transparency.

Note that the `alpha` command will apply to all commands that appear before it; thus you can control which objects get transparency by the order of the `alpha` and plotting commands, as shown in Fig. 7.9 below.

```
fill(xtriangle,ytriangle,'g');
hold on
fill(xcircle,ycircle,'r');
alpha(0.2)
```

```
fill(xsine,ysine,'b');
hold off
axis equal
```

You can also define different alpha to specific commands, or apply alpha to only one as the following code demonstrated in Fig. 7.10.

```
f1=fill(xtriangle,ytriangle,'g');
hold on
f2=fill(xcircle,ycircle,'r');
f3=fill(xsine,ysine,'b');
alpha(f3,0.4)
alpha(f2,0.7)
hold off
axis equal
```

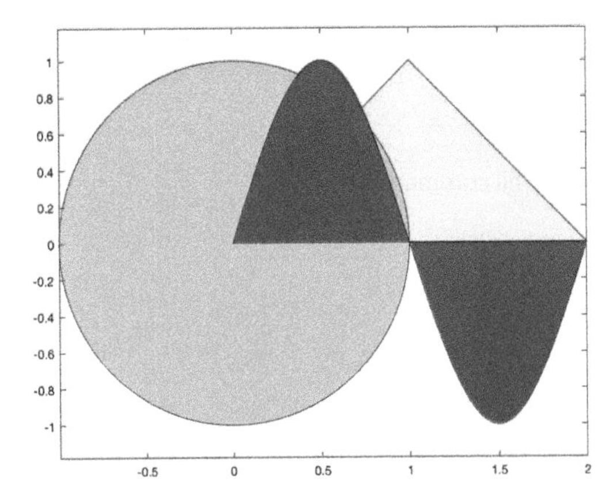

Figure 7.9 Order of `alpha` and other commands matter.

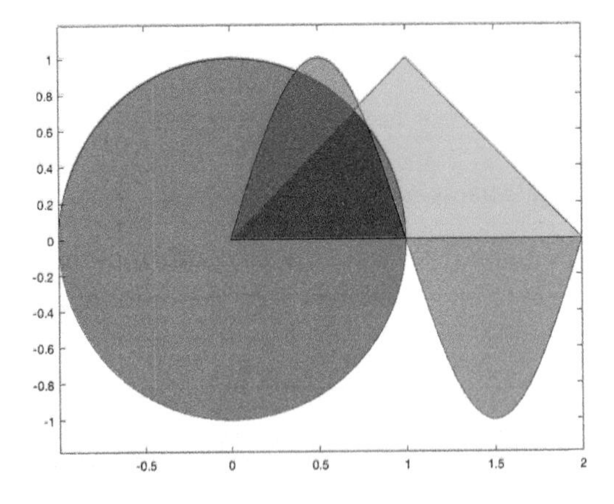

Figure 7.10 Order of `alpha` and other commands matter.

7.1.8 The syms, diff, int, and subs commands

The following subsection uses commands that require the Symbolic Math Toolbox. This may not be installed on your version of MATLAB.

There are many different commands and ways to create symbolic variables and functions within MATLAB. I am only giving a few examples here that are useful for some of the exercises.

```
>> syms x
>> f(x) = sin(pi*x)
f(x) =
sin(pi*x)
>> f(1)
ans =
0
>> A=f([0, 1/6, 1/4, 1/3, 1/2])
A =
[ 0, 1/2, 2^(1/2)/2, 3^(1/2)/2, 1]
>> double(A)        % this will give numeric approximations to A
ans =
         0    0.5000    0.7071    0.8660    1.0000
```

If you already have a string defined that we want to turn into a symbolic function, one can use the `str2sym` command for MATLAB R2017b or later. Otherwise one should use the `evalin` command. Other ways generate warnings and errors. Note that component-wise operations are not needed.

```
>> fstring = 'x^2 - 2*x + 1'
fstring =
    'x^2 - 2*x + 1'
>> syms x
>> f(x) = str2sym(fstring)
f(x) =
x^2 - 2*x + 1
>> f(3)
ans =
4
>> f([0:2])
ans =
[ 1, 0, 1]
>> f([0 1;2 3])
ans =
[ 1, 0]
[ 1, 4]
```

Symbolic variables can be declared and manipulated algebraically.

```
>> syms x y
>>  g(x)=x^3*sin(x);
>> t=linspace(-10,10);
>> plot(t,g(t))
>> expand((x+y)^3)
ans =
x^3 + 3*x^2*y + 3*x*y^2 + y^3
>> factor(x^2 + x*y -2*y^2)
ans =
[ x - y, x + 2*y]
```

The `diff` command can be used to differentiate a symbolic function.

```
>> syms x
>> f(x)=sin(pi*x) + exp(x);
>> g(x)=25*x^4;
>> df = diff(f)
df(x) =
exp(x) + pi*cos(pi*x)
>> dg = diff(g,x)
dg(x) =
100*x^3
>> dg(-1)
ans =
-100
>> syms t
>> h(x,t) = t^2 - 5*t*x + x^3
h(x, t) =
t^2 - 5*t*x + x^3
>> dhdt = diff(h,t)
dhdt(x, t) =
2*t - 5*x
```

Similarly, the `int` command can be used for integrals, both definite and indefinite.

```
>> f,g,h
f(x) =
sin(pi*x) + exp(x)
g(x) =
25*x^4
h(x, t) =
t^2 - 5*t*x + x^3
>> int(f)
ans(x) =
exp(x) - cos(pi*x)/pi
>> int(g,1,2)
ans =
155
```

```
>> inthdt = int(h,t)
inthdt(x, t) =
(t*(2*t^2 - 15*t*x + 6*x^3))/6
>> expand(inthdt)
ans(x, t) =
t^3/3 - (5*t^2*x)/2 + t*x^3
```

7.1.9 Commands for polynomials

Polynomials in MATLAB can be "stored" with a vector of their coefficients, with the **leading coefficient first and the constant term last**. Remember that in this case you must have 0s for the terms of that degree that are not a part of the polynomial.

Thus for the polynomial $p_1(x) = \pi x^2 - 2x + 3$ one can define within MATLAB the vector `p1=[pi -2 3];` and for the polynomial $p_2(x) = 4x + 1/2$ one can define `p2=[4 1/2]`. In order to work with both of them at the same time, the vectors should be the same size. Thus it would be better to define `p2=[0 4 1/2]`. We can easily add/subtract polynomials, since when we do this mathematically we just add/subtract the coefficients of like terms. When the coefficients are stored in vectors of the same size, this is equivalent to adding/subtracting the vectors.

```
>> p1=[pi -2 3];
>> p2=[0 4 1/2];
>> p1+p2
ans =
    3.1416    2.0000    3.5000
>> p1-4*p2
ans =
    3.1416  -18.0000    1.0000
```

The command `conv` (short for **convolution**) will do the appropriate algebra on vectors that is equivalent to multiplication of polynomials. Note that for `conv` the vectors do not need to be of the same size. The command `deconv` will perform polynomial division.

```
>> conv([3 2],[1 2])
ans =
     3     8     4
>> conv([1 -2], [1 2])
ans =
     1     0    -4
>> conv([1 -2 1], [1 -1])
ans =
     1    -3     3    -1
>> deconv([1 0 -9], [1, 3])
```

```
ans =
     1    -3
```

The `polyder` and `polyint` commands will return vectors that are the coefficients for the derivatives and integrals, respectively, of the polynomials/vectors.

```
>> polyder([3 2 -4 3])  % 3x^3 + 2x^2 -4x + 3
ans =
     9     4    -4
>> polyint([1 4 -3])     % x^2 + 2x  - 3
ans =
    0.3333    2.0000   -3.0000         0
```

The command `polyval` evaluates the given polynomial at the indicated value.

```
>> polyval([1 -2 3],3)  % x^2 - 2x + 3
ans =
     6
```

The `roots` command should be self-explanatory, but realize that it will also return complex roots.

```
>> sln1 = roots([1 2 -3])
sln1 =
   -3.0000
    1.0000
>> sln2 = roots([1 2 1])
sln2 =
    -1
    -1
>> sln3 = roots([1 0 1])
ans =
   0.0000 + 1.0000i
   0.0000 - 1.0000i
```

The command `polyfit(x,y,n)` will try and fit (using least-squares) the input data (x, y) to a polynomial of degree n as the following example demonstrates.

Example 7.1.1. Fit the points $(0, 0)$, $(-1, 1)$, and $(2, -2)$ to a line. State the line and plot the data points and line.

We can use `polyfit` with $n = 1$ for a line.

```
x=[0, -1, 2]; y= [0, 1, -1];
p=polyfit(x,y,1)
xplot=linspace(-2,3);
yplot=polyval(p,xplot);
plot(x,y,'*')
```

```
hold on
plot(xplot,yplot,'k')
hold off
```

The above code returns

```
p =
   -0.642857142857143    0.214285714285714
```

From this output, we see that the least-squares fit is

$$p(x) = (-0.642857142857143)x + 0.214285714285714,$$

and this line and these points are shown in Fig. 7.11.

Fig. 7.12 demonstrates the results of `polyfit` using polynomials of different degrees on `x=0.5; y=x.*cos(x).*exp(x).`

7.2. Code improvement

There are many ways one can improve code to make the programs run more efficiently and faster. One way is to reduce the number of unnecessary variables or steps. Sometimes, it is easier to write a program creating intermediate variables but then you can go back and put those calculations or comparisons within the same line. Another way is to fine tune the data types used in variables. For example, if you know that a variable will be an integer between −128 and 127, you can specify the variable to be stored as `int8` to save memory and execution time. These methods are often covered in a computer

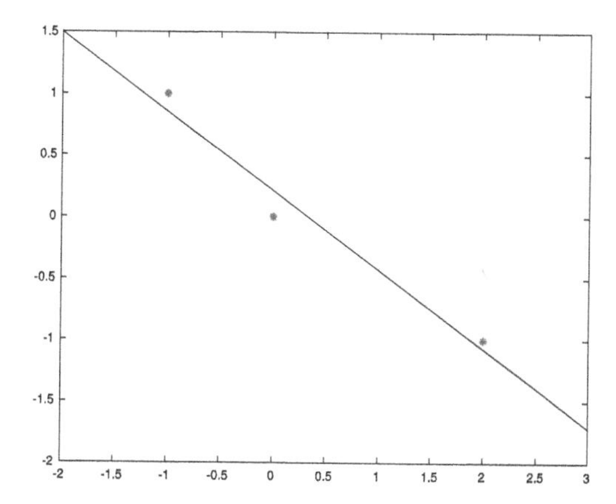

Figure 7.11 Polyfit example for line.

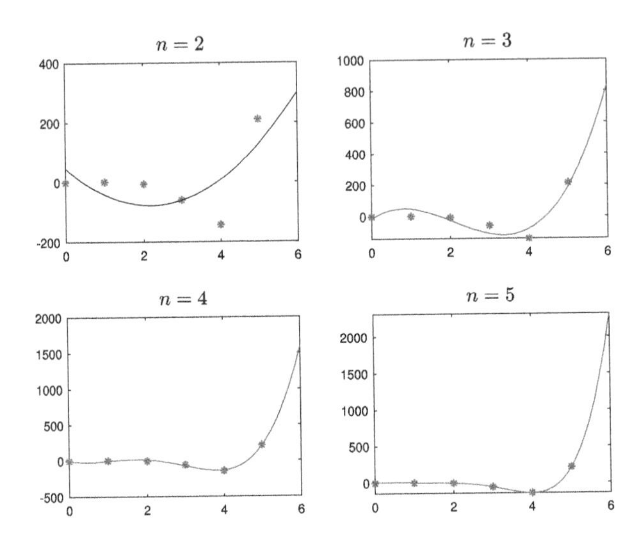

Figure 7.12 Polyfit example for different polynomial degrees.

science course and we will not go into details about this here. Many of our exercises in this course will not warrant this type of detail.

This section will focus on two basic methods of code improvement that may not be available in other languages: vectorization of code, and preallocation.

7.2.1 Vectorization of code

The method of vectorizing code is to capitalize on MATLAB's many features and commands with vectors and matrices. In many languages, a loop (usually a for loop) is needed to accomplish a task/operation on a vector, or over a set number of times. in MATLAB, sometimes one can create a vector and use the task/operation once on the entire vector. Note that there are situations in which you may still need to use loops on vectors and matrices. This may be explained best with an example.

Example 7.2.1. Calculate the geometric series the traditional way and using vectorization of code and compare using the `tic` and `toc` commands.

```
function [t1, t2] = geomseries2(r, n)
% GEOMSERIES2(R,N) uses the geometric series to demonstrate the vectorization of
    code.
%      [T1, T2] = GEOMSERIES2(R,N) will calculate two times using tic/toc.
%          T1 is the time taken to calculate the summation of
%               R^k from k = 0 to N using the traditional way.
%          T2 is the time taken to calculate the summation of
%               R^k from k = 0 to N using vectorization of code.
```

```
% Traditional way
    tic
    s1=0;
    for k=0:n
        s1=s1+r^k;
    end
    t1 = toc;
% Vectorization of code
    tic
    k=0:n;
    s2=sum(r.^k);
    t2 = toc;
end
```

Now look at the difference in times. For various values of n, we now run the following code.

```
[t1,t2]=geomseries2(0.5,n);
fprintf('\nThe two times for n = \%i are: \n\ttime1 = \%2.5f and \n\ttime2 = \%2.5f\n
    \n', n, t1, t2);
```

The output of the above code follows.

```

The two times for n = 10000 are:
        time1 = 0.01474 and
        time2 = 0.00045


The two times for n = 100000 are:
        time1 = 0.10631 and
        time2 = 0.00269


The two times for n = 1000000 are:
        time1 = 1.06955 and
        time2 = 0.02689

>>
```

7.2.2 Preallocation

One can make programs more efficient by preallocating the size of large vectors or matrices rather than appending. For example, if you have a recursive algorithm, you can preallocate a vector or matrix for the number of entries needed to complete the algorithm. Again, explanation will be by example.

Example 7.2.2. Demonstrate the efficiency of preallocation by calculating the first n Fibonacci numbers.

```
function F = Preallocation(n)
% PREALLOCATION(N) demonstrates the efficiency of preallocation.
%       PREALLOCATION(N) calculates the first N Fibonacci numbers.

% Traditional way.
F1=[1 1];
tic
for k=3:n
    nextOne = F1(k-1) + F1(k-2);
    F1 = [F1 nextOne];
end
toc
% Preallocation
tic
F2=ones(1,n);
for k=3:n
    F2(k) = F2(k-1) + F2(k-2);
end
toc
F = F2;
```

The comparison can be made by looking at the following output.

```
>> n=1000, Preallocation(n);
n =
        1000
Elapsed time is 0.002641 seconds.
Elapsed time is 0.000031 seconds.
>> n=10000, Preallocation(n);
n =
       10000
Elapsed time is 0.025771 seconds.
Elapsed time is 0.000274 seconds.
>> n=1000000, Preallocation(n);
n =
     1000000
Elapsed time is 2.020418 seconds.
Elapsed time is 0.018352 seconds.
```

PART 2

Mathematics and MATLAB®

CHAPTER 8

Transformations and Fern Fractals

8.1. Linear transformations

Linear transformations have many special properties in addition to the properties from the definition below. Linear transformations preserve lines, as opposed to nonlinear transformations that may transform a line segment into a parabolic shape, or an ellipse, etc.

> **Definition 8.1.1. Linear transformations** are functions or mappings from a vector space V to a vector space W are transformations (functions) $T : V \to W$ such that
> - $T(\mathbf{u}+\mathbf{v}) = T(\mathbf{u}) + T(\mathbf{v})$ for all $\mathbf{u}, \mathbf{v} \in V$.
> - $T(c\mathbf{u}) = cT(\mathbf{u})$ for all $\mathbf{u} \in V$, $c \in \mathbb{R}$.

It is discussed in linear algebra that linear transformations can be associated (uniquely) with matrices. We will be looking at linear transformations from $\mathbb{R}^2$ to $\mathbb{R}^2$, so these transformations correspond to 2×2 matrices (with $\mathbb{R}$-entries). Thus $T(\mathbf{x}) \leftrightarrow A\mathbf{x}$, where A is a 2×2 matrix and $\mathbf{x}$ is a 2×1 matrix or column vector.

Special examples of linear transformations are rotations, translations, and scaling transformations (dilations/contractions).

- **Scaling transformation** (see Fig. 8.1) is a transformation $T : \mathbb{R}^2 \to \mathbb{R}^2$ defined by $T(\mathbf{x}) = c\mathbf{x}$ where $c \in (0, \infty)$. Using properties of scalar multiplication on vectors, we see that the image is now c times longer (or shorter) than the input but is in the same direction. Thus
  - If $c > 1$, T is a *dilation by a factor of* c.
  - If $c \in (0, 1)$, T is a *contraction by a factor of* c.

  We can put T into matrix form:

$$T\left(\begin{bmatrix} x \\ y \end{bmatrix}\right) = \begin{bmatrix} c & 0 \\ 0 & c \end{bmatrix}\begin{bmatrix} x \\ y \end{bmatrix} = \begin{bmatrix} cx \\ cy \end{bmatrix}$$

- **Rotation** by an angle θ about the origin $\mathcal{O}$ where the rotation is measured from the positive x-axis (see Fig. 8.2) can be represented as:

$$T\left(\begin{bmatrix} x_1 \\ x_2 \end{bmatrix}\right) = \begin{bmatrix} \cos\theta & -\sin\theta \\ \sin\theta & \cos\theta \end{bmatrix}\begin{bmatrix} x_1 \\ x_2 \end{bmatrix}.$$

Programming Mathematics Using MATLAB®
https://doi.org/10.1016/B978-0-12-817799-0.00014-4

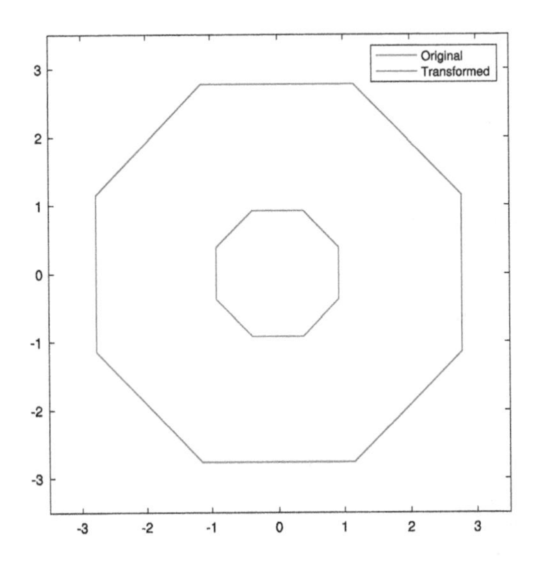

Figure 8.1 Scaling example.

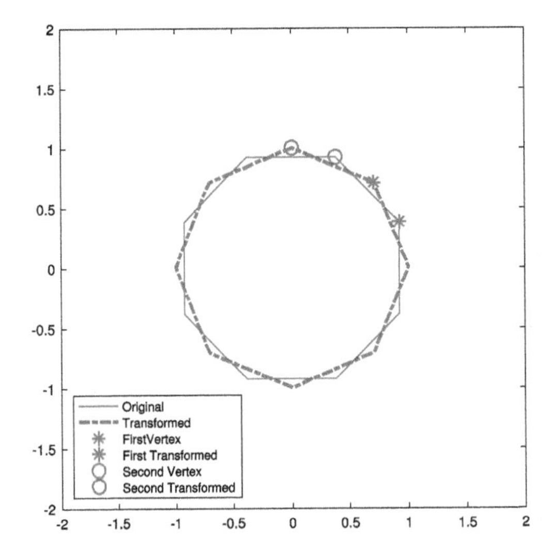

Figure 8.2 Rotation example with $\theta = \frac{\pi}{8}$.

- **Reflection** about a line ℓ through the origin, where θ is the angle from the positive x-axis to ℓ is given by

$$A = \begin{bmatrix} \cos 2\theta & \sin 2\theta \\ \sin 2\theta & -\cos 2\theta \end{bmatrix} = \begin{bmatrix} a & b \\ b & -a \end{bmatrix}, \quad a^2 + b^2 = 1.$$

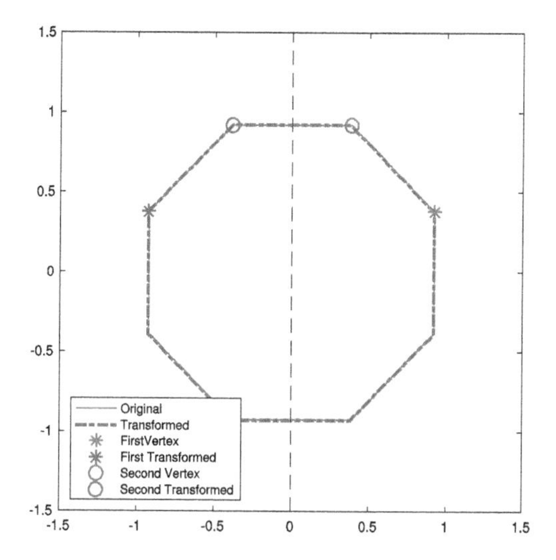

Figure 8.3 Reflection example about the y-axis.

Example 8.1.1. $T\left(\begin{bmatrix} x \\ y \end{bmatrix}\right) = \begin{bmatrix} -x \\ y \end{bmatrix}$ reflects the point (vector) in $\mathbb{R}^2$ about the y-axis (see Fig. 8.3). Using $\theta = \pi/2$, the matrix for this transformation is given by the equation

$$T\left(\begin{bmatrix} x \\ y \end{bmatrix}\right) = \begin{bmatrix} -1 & 0 \\ 0 & 1 \end{bmatrix}\begin{bmatrix} x \\ y \end{bmatrix}$$

- **Composition of transformations** Consider two linear transformations T and S, both from $\mathbb{R}^2$ to $\mathbb{R}^2$. Suppose A is the matrix corresponding to T and B is the matrix corresponding to S; i.e.,

  $$T(\mathbf{x}) = A\mathbf{x} \qquad \text{and} \qquad S(\mathbf{x}) = B\mathbf{x} \quad \forall\, \mathbf{x} \in \mathbb{R}^2$$

  Then the composition of the transformations T and S, $T \circ S$, is given by the matrix AB;

  $$(T \circ S)(\mathbf{x}) = T(S(\mathbf{x})) = T(B\mathbf{x}) = A(B\mathbf{x}) = AB\mathbf{x}.$$

If you think about some transformations that are performed after each other, the order in which they are done will matter. Likewise, since matrix multiplication is not commutative, the order matters (see Fig. 8.4).

The above are examples in which the matrices have particular patterns or formats to generate particular transformations or geometric results. Any matrix will yield a

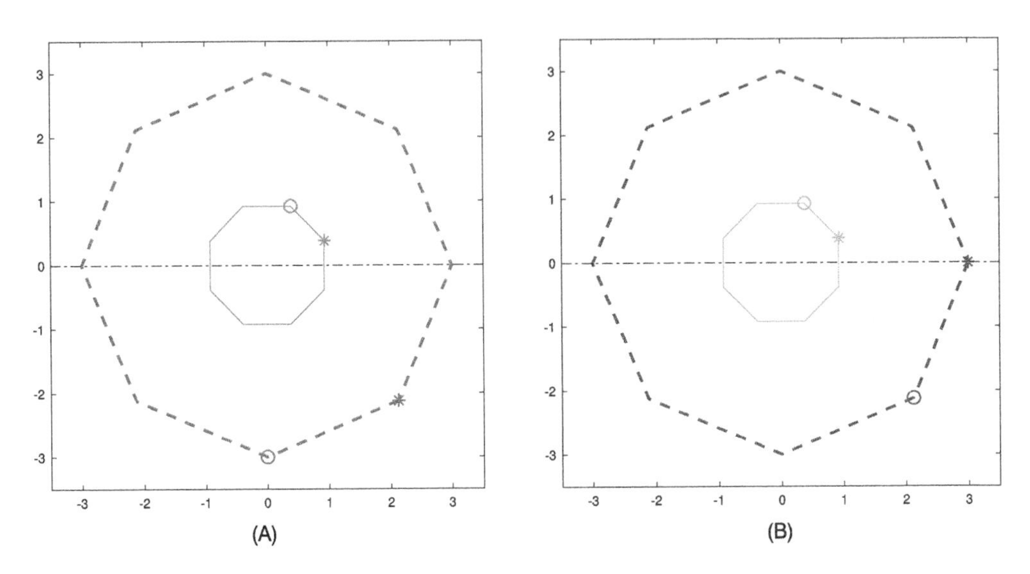

Figure 8.4 Composition example. (A) Rotation, then reflection, then dilation, (B) Dilation, then reflection, then rotation.

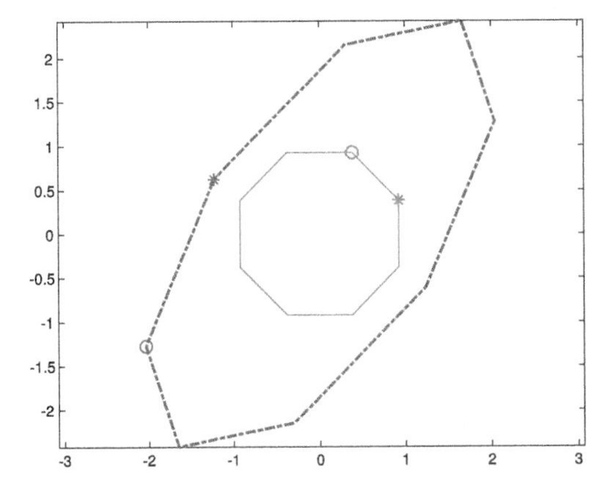

Figure 8.5 Random linear transformation.

linear transformation; the big difference being when transforming a polygon as in our examples, the angles between sides are not necessarily preserved and lengths of the sides may not be scaled equally. For example, consider the linear transformation obtained from the matrix M (see Fig. 8.5):

$$M = \begin{bmatrix} -\frac{1}{2} & -2 \\ \frac{3}{2} & -2 \end{bmatrix}$$

The reflection and rotation transformations are examples of special transformations known as **orthogonal transformations**. Orthogonal transformations are linear transformations $T : \mathbb{R}^n \to \mathbb{R}^n$ such that

$$\|T(\mathbf{x})\| = \|\mathbf{x}\|.$$

These transformations preserve the lengths of the vectors and the inner or dot products. Thus in Euclidean vector spaces, the angles between vectors are preserved in addition to the lengths of the vectors. This means that under these linear transformations, right triangles transform to right triangles of the same size. Contraction and dilation transformations also preserve angles, but the lengths of the vectors change so right triangles transform to *similar* right triangles. Other transformations will transform right triangles to triangles, but the transformed triangles are not necessarily right triangles and the sides are not necessarily the same length or even same proportion.

8.2. Affine transformations

Affine transformations are not exactly linear transformations, but are similar in that they can be written in terms of matrices and vectors.

Definition 8.2.1. Affine transformations are functions or mappings from one vector space V to another vector space W of the form

$$T(\mathbf{x}) = M\mathbf{x} + \mathbf{v}$$

where M is a given matrix and $\mathbf{v}$ is a given vector.

One can think of affine transformations as linear transformations ($M\mathbf{x}$) then shifted by the vector $\mathbf{v}$. Similar to linear transformations, affine transformations preserve straight lines: thus triangles remain triangles, quadrilaterals remain quadrilaterals, etc.

The example below shows examples of two affine transformations.

Example 8.2.1. The first affine transformation shown in Fig. 8.6 is

$$\begin{bmatrix} \frac{1}{3} & 0 \\ 0 & -\frac{1}{3} \end{bmatrix} \mathbf{x} + \begin{bmatrix} 3 \\ 4 \end{bmatrix}.$$

The second affine transformation shown in Fig. 8.6 is

$$\begin{bmatrix} 1 & \frac{1}{4} \\ \frac{1}{4} & \frac{3}{4} \end{bmatrix} \mathbf{x} + \begin{bmatrix} 3 \\ 4 \end{bmatrix}.$$

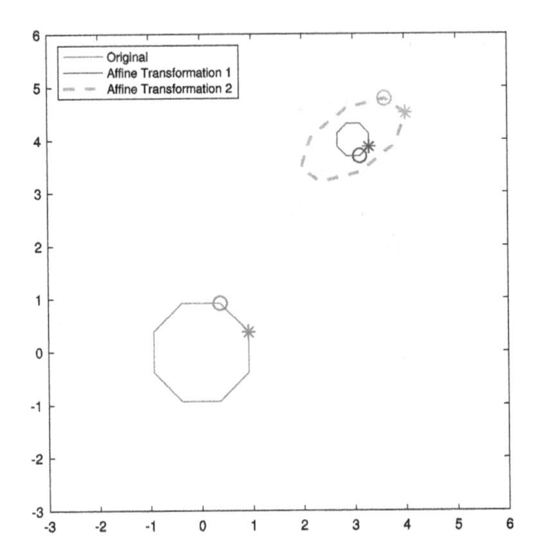

Figure 8.6 Affine transformation examples.

8.3. Fern fractals

"Clouds are not spheres, mountains are not cones, coastlines are not circles, and bark is not smooth, nor does lightning travel in a straight line" – Benoit Mandelbrot [16]

According to many stories, Mandelbrot started thinking about "fractals" when trying to measure the coastline of Britain. He coined the term "fractal" from Latin "fractus" meaning "broken" or "fractured".

Fractals have two main distinctions:

1. Self-similarity (pattern repeats itself at magnification).
2. Non-integer dimension (Hausdorff dimension) which will not be discussed at this time.

Fractals can involve many different areas of mathematics such as complex analysis, geometry, dynamical systems, linear algebra and the visualization of them is considered by many to be beautiful.

Fern fractals are examples of **iterated function systems** (IFSs) and are generated by affine transformations. We will be looking at the Barnsley fern fractal [2], [19]. The basic premise of basic fern fractals is as follows:

- Four affine transformations each have a probability of being chosen.
  1. The transformation with the highest probability generates successively smaller leaflets.
  2. Two transformations have equal probabilities. One of them generates the largest left-hand leaflet, the other generates the largest right-hand leaflet.
  3. The transformation with the smallest probability generates the stem.

- Start at the origin, one of the four transformations is chosen based on its probability. The next point is calculated based on this transformation calculated on the current point.
- Repeat this process with each new point calculated.

For a specific IFS, the image created will resemble ferns. In order to work with the probabilities and chosen transformations, we will split up the interval [0, 1] based on probabilities. For example, if there are three transformations, the first having probability 0.25, the second 0.5, and the third 0.25, then we will select which transformation is chosen by using `rand` to generate a number between 0 and 1. If the number is between 0 and 0.25, the first transformation is selected. If the number is between 0.25 and 0.75, the second transformation is selected. Lastly, if the number is between 0.75 and 1, the third transformation is selected. Which transformations get the endpoints of 0.25 and 0.75 does not really matter.

8.4. Exercises

1. **Set up**
   (a) We will define the RGB code for four colors are as follows:

Name	RGB Code	Variable name
My Green	0-104-87	mygreen
My Gray	200-200-200	mygray
My Gold	250-227-125	mygold
My Red	179-8-56	myred

   Save the `MyColors.mat` file found in on the text website to your folder/directory and use the `load('MyColors')` command to load the vectors that can be used as colors.

   (b) One of the shapes we will use in these exercises will be an ellipse made with 100 points using the parametric equations $x = 2\cos(t)$ and $y = 3\sin(t)$ for an appropriate domain for t. Set up the vectors of these values (variable names of your choice!) to plot the ellipse in this exercise and in subsequent exercises. To test, use the `plot` command to plot the ellipse in mygreen with a Line Width of 2.

2. **Visualizing linear transformations**
   For this exercise we will be visualizing some examples of linear transformations. The process will be the same or similar for all of them, so copy/paste will be your friend. Remember that using the `plot(x,y)` command, Matlab® "connects the dots" of the points given with `x` and `y` as vectors of the x- and y-coordinates of the points. We will use this idea to draw shapes and the transformed shapes. One of the shapes will be the ellipse discussed in #1b. Note that once the x and y vectors for

the ellipse are created in #1b, as long as those vector names are unique you do not need to recreate them in subsequent exercises. The other shapes will be made using your function `myCircle2` that is a (fixed) copy of the function from Exercise 9 in Chapter 6.

Our shapes will be plotted by forming a vector comprised of the x-values and another comprised of the y-values of the points that make up our shapes. Each of our transformations will be associated with a 2×2 matrix A. What the matrices should be for reflections, rotations, scaling, and compositions are in the lecture/class notes. To visualize the transformation associated with a matrix A, we want to calculate $A\mathbf{v}$ for each of the points $\mathbf{v}$ (which is actually a column vector) that make up our shapes and store the answer in a new set of points. But this would be tedious and/or inefficient to do this for each point. Instead, we can do this all at once to all of the points by using matrix calculations. This is where it is easiest to have the original points in a $2 \times n$ matrix V (n is determined by the number of points that make up our shape). If we set up the matrix V so that the first row contains the x-values of the points, and the second row contains the y-values, then each *column* of V is one point. The beauty of matrix multiplication is that $T = AV$ will be all of the points that make up the transformed shape. For each of the linear transformations we will do the following.

- Formulate the vectors `x`, `y`, and matrix `V` that are composed of the vertices or points of the specified original shape. Plot the original shape in mygreen with Line Width thicker than the default.
- Figure out what the appropriate matrix A would be for the transformation, and display that matrix. In other words, when defining the matrix for each transformation, do not suppress the output so we can see the values in the matrix when it is published. It may be better to have a unique name for each transformation matrix, instead of A for each problem.
- Calculate $T = AV$ and plot the transformed shape in the stated color with Line Width thicker than the default.
- Plot the two stated points (markers) of the original shape in mygreen, using the "*" marker and square marker, respectively. Plot the two transformed markers using the same shapes, but in the same color as the transformed shape. Thus you should be able to see where the markers went under the transformation.
- Use the command `axis equal` and set the axes appropriately so the edges and points of the shapes do not touch the edge of the figure. **You may have to experiment with the order of `axis equal` and the command(s) to set the axes.**
- Use a legend with the titles "Original" and "Transformed" for the 2 shapes, and "1st marker" and "2nd marker" for the specified points. Note that the order of your plot commands will affect the order the titles should be in the

legend! You may want to change the location of the legend with `'Location', 'Southeast'` (or some other location specification) so you can see all of the figures and legend. (This location may need to change for different transformations below.) For example,

```
legend('Sample title1', 'Sample title2', 'Location', 'Southeast').
```

Perform the above process for each of the following transformations and shapes. Have the text that is in bold be the title of your figure (first example, `title('Reflection Example'))`

(a) **Reflection example** with $\theta = \pi/6$. Original shape: the ellipse. Transformed shape in myred. First marker is the first point, the second marker is the 25th point. Also draw the reflection line of the angle θ (easiest to use trig!) as a black, dashed line.

(b) **Scaling examples** with $c = 1.5$ and $c = 1/2$. Original shape: a hexagon using your `circle2` function. Transformed shapes are in mygray and mygold, respectively. First marker is the first point, the second marker is the second point. Plot all three shapes: the original and two transformed, on the same figure.

(c) **Rotation example** with $\theta =?$. Original shape: a pentagon. Notice that using your `circle2` function to create the pentagon makes the shape "off kilter" or tipped. We want to orient the pentagon so the bottom side is perfectly horizontal, so we need to rotate it by an appropriate angle θ. Figure out what θ should equal and use this angle for your rotation example. Transformed shape should be in myred. First marker is the first point, the second marker is the second point.

(d) **Two composition examples** Original shape: a triangle using your `circle2`. Transformed shapes are in myred and mygold, respectively. First marker is the first point, the second marker is the second point. The first transformation is a reflection about the y-axis (what is θ?) followed by a rotation with $\theta = \pi/4$. The second transformation is the rotation and then reflection. Plot all three shapes on the same figure.

(e) **Two random transformations** Original shape: the ellipse. Transformed shapes are in mygray and mygold. First marker is the first point, the second marker is the tenth point. The matrix for these random linear transformations are 2×2 matrices with random integer entries from -5 to 5.

(f) **Two random transformations** Original shape: the square. Transformed shapes are in mygray and mygold. First marker is the first point, the second marker is the second point. The matrix for these random linear transformations are 2×2 matrices with random integer entries from -5 to 5.

3. **Visualizing affine transformations** For this problem we will be doing the following:

- Save the `affineData.mat` file found on the text website to your folder/directory and use the `load('affineData')` command to load the matrices and vectors for the affine transformations (listed below).
- Draw the hexagon using your `circle2` in black.
- Calculate the transformed hexagon using the affine transformations below (T_1 for the first transformation, T_2 for the second, etc.).
- Plot the transformed hexagons in mygreen, myred, and mygold, respectively.
- Use the command `axis equal` and define the axes to an appropriate view so the hexagons do not touch the edge of the figure.
- Have a legend identifying the original, first transformation, etc.
- No axis labels are needed for these graphs.
- The title of the graph should be "Affine Transformations."

Affine transformations $T_k(\mathbf{x}) = M_k\mathbf{x} + \mathbf{v}_k$, $(k = 1, 2, 3)$ where

$$M_1 = \begin{bmatrix} 0 & 1 \\ 3 & 1 \end{bmatrix}, \ \mathbf{v}_1 = \begin{bmatrix} 1.5 \\ 0 \end{bmatrix}, \quad M_2 = \begin{bmatrix} 3 & 1 \\ 0 & 1 \end{bmatrix}, \ \mathbf{v}_2 = \begin{bmatrix} -1.25 \\ -1.25 \end{bmatrix},$$

$$M_3 = \begin{bmatrix} -3 & 0 \\ 2 & 3 \end{bmatrix}, \ \mathbf{v}_3 = \begin{bmatrix} -1 \\ 1 \end{bmatrix}.$$

(NOTE: your points/vectors $\mathbf{x}$ should be a column vectors for this notation).

4. **Fern fractals** We will create functions `fern1`, `fern2`, and `fern3` that have as input a positive integer n and optional input variables. The function should check for the validity of n, if not, return an error. You may use your previously written functions to help with this error check. Optional input variables will be plot specifiers (color, marker size, marker type, etc.). The functions will use the affine transformations defined by the *iterated function system* (IFS) below using n iterations to generate a figure of a fern fractal. The output of the function will be the value from the `toc` command. Each affine transformation T_k $(k = 1, 2, 3, 4)$ involves the six parameters a, b, c, d, e, f:

$$T_k(\mathbf{x}) = M_k\mathbf{x} + \mathbf{v}_k = \begin{bmatrix} a & b \\ c & d \end{bmatrix} \mathbf{x} + \begin{bmatrix} e \\ f \end{bmatrix}$$

where each transformation has probability p_k of being performed. The IFS for the Barnsley fern [2, p. 86] we will be plotting is in Table 8.1.

The three functions will use slightly different algorithms. The first one is a basic one in which you may find the algorithm in a basic linear algebra text [27] or on Wikipedia for plotting these fractals in other languages.

(a) For `fern1`, here is the algorithm:

i. Use the command `tic`.

Table 8.1 IFS for Barnsley fern fractal.

T	a	b	c	d	e	f	p
1	0	0	0	$\frac{4}{25}$	0	0	$\frac{1}{100}$
2	$\frac{17}{20}$	$\frac{1}{25}$	$-\frac{1}{25}$	$\frac{17}{20}$	0	$\frac{8}{5}$	$\frac{17}{20}$
3	$\frac{1}{5}$	$-\frac{13}{50}$	$\frac{23}{100}$	$\frac{11}{50}$	0	$\frac{8}{5}$	$\frac{7}{100}$
4	$-\frac{3}{20}$	$\frac{7}{25}$	$\frac{13}{50}$	$\frac{6}{25}$	0	$\frac{11}{25}$	$\frac{7}{100}$

ii. Load the variables found in `fernIFS.mat` that can be found on the text website to have the matrices, vectors, and probabilities that are needed to do the following tasks.

iii. Let $\mathbf{x} = \begin{bmatrix} 0 \\ 0 \end{bmatrix}$.

iv. If n is the only input, plot the point $\mathbf{x}$ as a point using "$\star$" marker. Otherwise, plot the point with the additional inputs using `varargin`.

v. Use the random number generator `rand` to select one of the affine transformations T_k according to the given probabilities.

vi. Redefine $\mathbf{x}$ to be the new $\mathbf{x} = T_k(\mathbf{x}) = M_k\mathbf{x} + \mathbf{v}_k$.

vii. Plot the new point $\mathbf{x}$ as you did with the original $\mathbf{x}$.

viii. Repeat steps (vi) through (viii) so a total of n new points are plotted (where should the `hold on` and `hold off` commands be? How many times should your loop iterate?).

ix. The last command of the function should be storing `toc` into your output variable to capture the time it took to generate and plot the fern fractal.

(b) For `fern2`, we will adjust the process of `fern1` by **vectorizing the code**.

i. Use the command `tic`.

ii. Load the variables in `fernIFS.mat` found on the text website to have the matrices, vectors, and probabilities that are needed to do the following tasks.

iii. Let $\mathbf{x} = \begin{bmatrix} 0 \\ 0 \end{bmatrix}$ and set $X = \mathbf{x}$.

iv. Use the random number generator `rand` to create a vector of n random numbers to determine all of the affine transformations.

v. Repeat the following so a total of n new points are in the matrix X:

A. Select one of the affine transformations T_k according to the given probabilities using the appropriate element from the vector of random numbers created above.

B. Redefine $\mathbf{x} = T_k(\mathbf{x}) = M_k\mathbf{x} + \mathbf{v}_k$. Concatenate X with the new $\mathbf{x}$ (add another column to X that is the new $\mathbf{x}$).

vi. If n is the only input, plot the points in the matrix X as a point using "★" marker. Otherwise, plot the points with the additional inputs using `varargin`. (Note that the first row of X contains the x-coordinates, the second row contains the y-coordinates.)

vii. The last command of the function should be storing `toc` into your output variable to capture the time it took to generate and plot the fern fractal.

(c) For `fern3`, we will adjust the process of `fern2` by **preallocation**.

i. Use the command `tic`.

ii. Load the variables in `fernIFS.mat` found on the text website to have the matrices, vectors, and probabilities that are needed to do the following tasks.

iii. Define $\mathbf{x}$ as a matrix with two rows and $n + 1$ columns of zeros.

iv. Use the random number generator `rand` to create a vector of n random numbers to determine all of the affine transformations.

v. For columns 2 through $n + 1$ of $\mathbf{x}$, use the appropriate element from the vector of random numbers created above to select one of the affine transformations T_k according to the given probabilities and let that column of $\mathbf{x}$ equal T_k(previous column of $\mathbf{x}$).

vi. If n is the only input, plot the points in the matrix $\mathbf{x}$ as a point using "★" marker. Otherwise, plot the points with the additional inputs using `varargin`. (Note that the first row of $\mathbf{x}$ contains the x-coordinates, the second row contains the y-coordinates.)

vii. The last command of the function should be storing `toc` into your output variable to capture the time it took to generate and plot the fern fractal.

5. Display the fern using `fern1`, `fern2`, and `fern3` with $n = 2000$. Before each fern command, use the command `clf`, and after each fern command, use the commands `axis equal` and `axis off`. For each of the ferns, use the `sprintf` command to create titles that say which fern function, the value of n, and how much time elapsed to create the fern fractal.

6. Display the fern using `fern1`, `fern2`, and `fern3` with $n = 5000$, and specify the square marker, color mygreen, and have the marker size be smaller than the default. Before each fern command, use the command `clf`, and after each fern command, use the commands `axis equal` and `axis off`. For each of the ferns, use the `sprintf` command to create titles that say which fern function, the value of n, and how much time elapsed to create the fern fractal.

7. Display the fern using `fern2` and `fern3` with $n = 50{,}000$ and specify a marker and a color of your choice (do not use the default color). Use other specifiers if you want. Before each fern command, use the command `clf`, and after each fern command, use the commands `axis equal` and `axis off`. For each of the ferns, use the `sprintf` command to create titles that say which fern function, the value of n, and how much time elapsed to create the fern fractal.

CHAPTER 9

Complex Numbers and Fractals

9.1. Complex numbers

Complex numbers ($\mathbb{C}$) are actually points in $\mathbb{R}^2$ with additional structure. The point (a, b) in $\mathbb{R}^2$ would be the point $z = a + bi$ in $\mathbb{C}$. The a is the **real part** or z, and b is the **imaginary part** of z. The x-axis is now called the **real axis**, and the y-axis is now called the **imaginary axis** (see Fig. 9.1).

9.1.1 Adding complex numbers

As in $\mathbb{R}^2$, one can add and subtract these numbers "element-wise". The result is the same as adding vectors (using triangle or parallelogram law) (see Fig. 9.2).

9.1.2 Multiplication by a real numbers (scalars)

Just as in vectors, multiplying a complex number by a scalar (real number) c changes the norm/magnitude/length by a factor or c. Recall that multiplying by a negative number makes the vector point in the opposite direction (see Fig. 9.3).

9.1.3 Multiplication and de Moivre's theorem/formula

The biggest difference between vectors in $\mathbb{R}^2$ and complex numbers is the ability to multiply. We learned that there is no way to multiply vectors. But we can multiply complex numbers. If $z = a + bi$ and $w = c + di$, then to find zw we just "foil it out" and

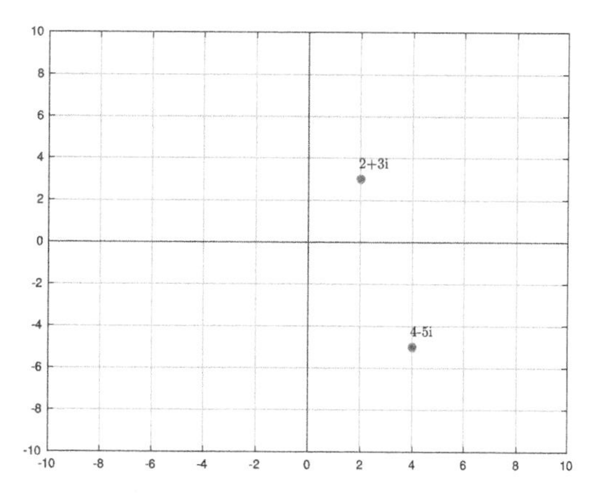

Figure 9.1 Complex numbers.

Programming Mathematics Using MATLAB®
https://doi.org/10.1016/B978-0-12-817799-0.00015-6

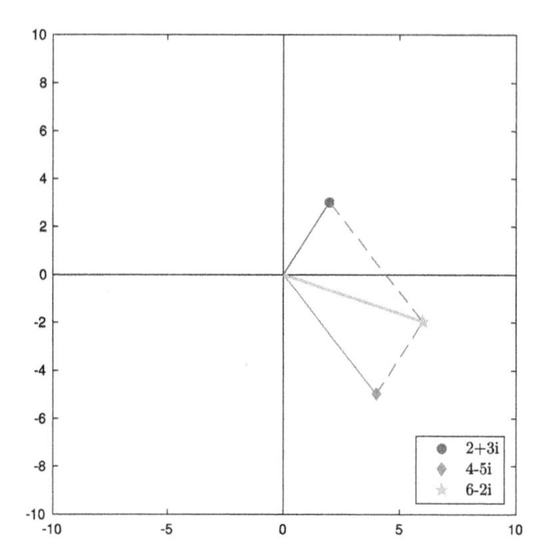

Figure 9.2 Adding complex numbers.

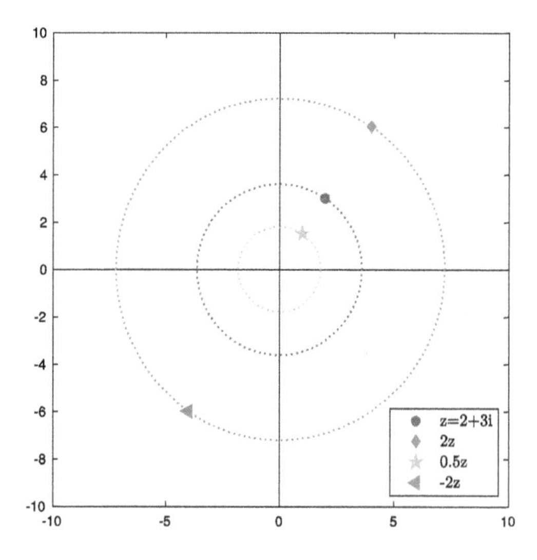

Figure 9.3 Multiplying complex numbers by a scalar.

remember that $i^2 = -1$ to get

$$zw = (ac - bd) + i(ad + bc).$$

This still does not show us how to understand the importance of using complex numbers for multiplication rather than vectors. Representing complex numbers in a different way shows us the meaning and use of multiplication.

As we learned in Calculus II, there are two ways one can identify points on the 2D plane: Cartesian coordinates (x, y) and polar coordinates (r, θ) using the conversion formulas:

$$\begin{aligned} x &= r\cos\theta, \\ y &= r\sin\theta, \\ x^2 + y^2 &= r^2, \\ \frac{y}{x} &= \tan\theta. \end{aligned}$$

With this idea, using $z = x + iy$ one has the **polar representation of a complex number**:

$$z = x + iy = r\cos\theta + ir\sin\theta$$

where r is the **length**, **norm**, or **magnitude** of z ($r = |z| = \sqrt{x^2 + y^2}$) and θ is the **argument** of z (the angle of rotation from the positive real-axis to z).

Using trigonometry and polar representations, we can visualize what happens when we multiply complex numbers. Using trigonometric identities, one can prove **de Moivre's formula/theorem**:

> **Theorem 9.1.1** (de Moivre's theorem)**.** When $z = r\cos\theta + ir\sin\theta$, and n is any natural number,
>
> $$z^n = r^n\cos(n\theta) + ir^n\sin(n\theta).$$

This means that when we take a complex number and raise it to a positive integer power n, the result is a complex number in which the length is taken to that power n and the argument is multiplied n times (thus rotated n times).

Consider the graph below. For the number $z = 2 + 2i$, we have $r = |z| = 2\sqrt{2}$, and $\theta = \frac{\pi}{4}$. Thus z^2 has length $r^2 = 8$, and argument $2\theta = \frac{\pi}{4}$. Also, z^3 has length $r^3 = 8\sqrt{8}$, and argument $3\theta = \frac{3\pi}{4}$ (see Fig. 9.4).

If we let $z = r\cos\theta + i\sin\theta$ and $w = s\cos\psi + is\sin\psi$, then using trigonometric identities we get

$$zw = rs\cos(\theta + \psi) + irs\sin(\theta + \psi).$$

Thus the **lengths are multiplied and the arguments (angles of rotation) are added**.

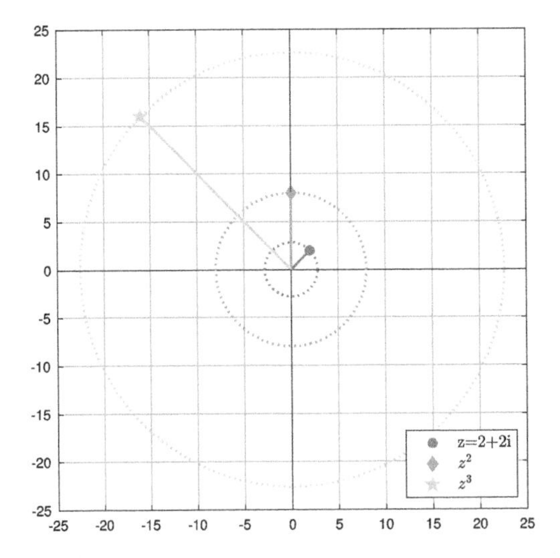

Figure 9.4 Powers of complex numbers.

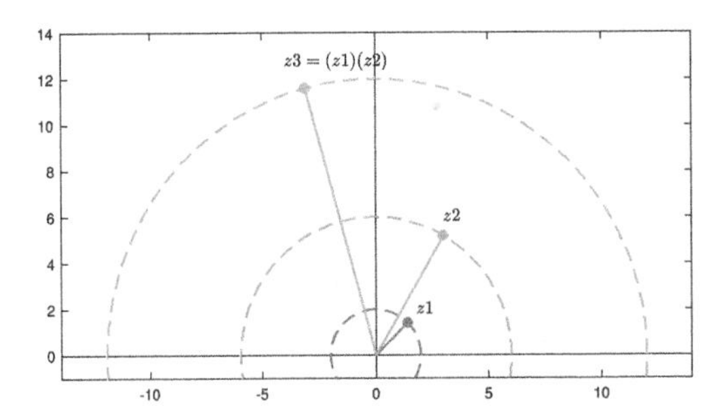

Figure 9.5 Multiplying complex numbers.

Consider the complex numbers pictured in Fig. 9.5: $z1 = 2\cos(\pi/4) + i2\sin(\pi/4) = \sqrt{2} + i\sqrt{2}$ and $z2 = 6\cos(\pi/3) + i6\sin(\pi/3) = 3 + (3\sqrt{3})i$. If we multiply them using the polar representation, we get

$$z3 = 12\cos(7\pi/12) + i12\sin(7\pi/12). = (3\sqrt{3} - 3\sqrt{6}) + i(3\sqrt{6} + 3\sqrt{2}) \approx -3.1058 + 11.5911i.$$

9.1.4 Plotting complex numbers in MATLAB®

One nice thing about complex numbers is it makes plotting points "easier" in MATLAB. Say we wanted to plot and/or connect the two points $(2, 3)$ and $(6, 5)$. As points in $\mathbb{R}^2$, we would have to do something like the following:

```
x1=[2;3]; x2=[6;5];
x=[x1 x2];
plot(x(1,:),x(2,:),'*')
```

Using complex numbers instead, the same picture could be created by

```
a=2+3i; b=6+5i;
plot([a b],'*')
```

Similarly, if we wanted to connect the points, we can do the following:

```
a=2+3i; b=6+5i;
plot([a b])
```

A word of caution: if you have a purely real number and want it plotted as a complex number, then you either have to have it within a vector of other complex numbers or plot the real and imaginary part separately. For example, `x=[2,i]; plot(x,'*')` will plot the 2 correctly, but if you have `x=2;` instead, or even `x=2+0*i`, you will not get the desired results.

9.1.5 Creating line segments with complex numbers

If we wanted to look at the line segment (vector) connecting points $A(2,3)$ and $B(6,5)$, in multivariable calculus we learn that the coordinates of the **position vector** (vector originating at the origin) will be the coordinates of B minus the coordinates of A. In our example, it would have the coordinates $(4,2)$. If instead we take the complex number version of the points $a=2+3i$, and $b=6+5i$, then if we look at $b-a$ we get $b-a=4+2i$. This vector, whether it is written in $\mathbb{R}^2$ or in $\mathbb{C}$, has the same length and direction as the vector connecting point A to B, as shown in Fig. 9.6.

If we wanted to move that line segment to start at another point, say, $1+i$, the line segment would be between the points $1+i$ and $b-a+1+i$. If instead we wanted to take the vector $b-a$ and scale it by, say, $1/3$, then you just create the vector $\frac{b-a}{3}$ (see Fig. 9.7).

We could have this scaled vector start at a, or we could have this scaled vector start at b (see Fig. 9.8).

We can take that scaled vector and rotate it. Recall above that multiplying two complex numbers will add the arguments (angles of rotation in polar representation) and their lengths will be multiplied (see Section 9.1.3). Suppose we were to take the angle $\theta=-\frac{\pi}{2}$ and create a complex number $z=\cos\theta+i\sin\theta$. What would this number z equal? The length of z is 1, so any point that we multiply z by will have the same length, but the angle will be rotated by $\theta=-\frac{\pi}{2}$. This is shown in Fig. 9.9.

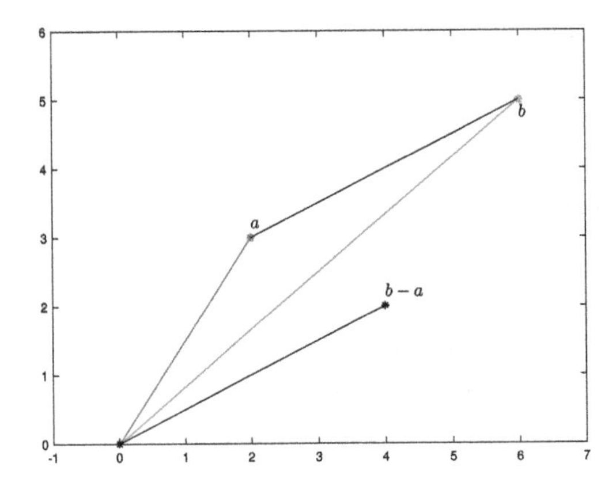

Figure 9.6 Line segment $b - a$.

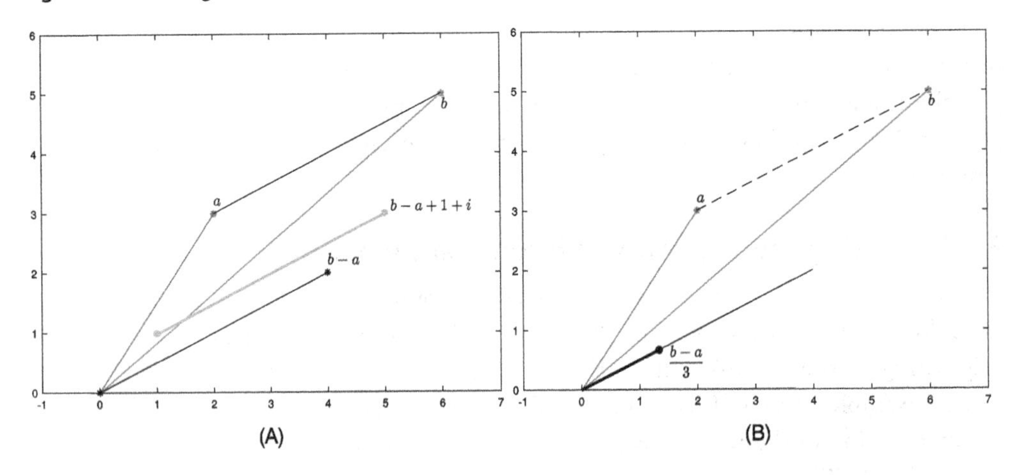

Figure 9.7 Scaling and shifting. (A) Shifting the line segment, (B) Scaling a line segment.

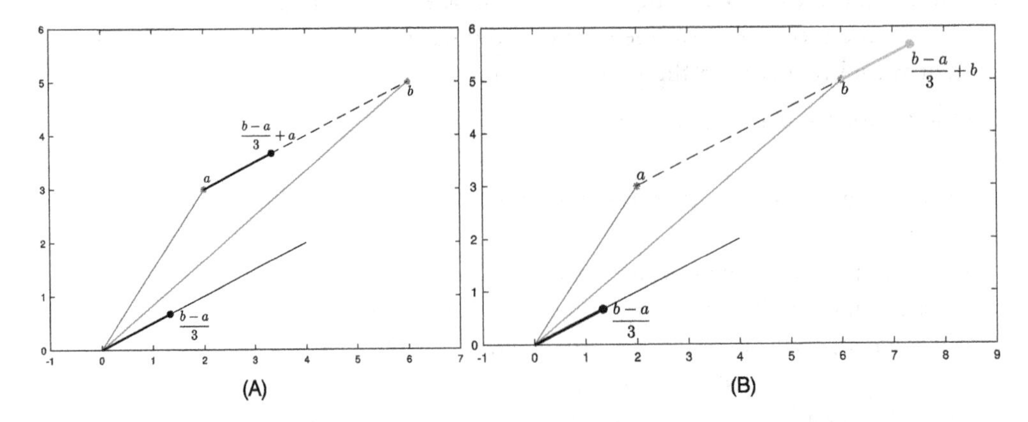

Figure 9.8 Scaling and shifting again. (A) Scaled vector starting at a, (B) Scaled vector starting at b.

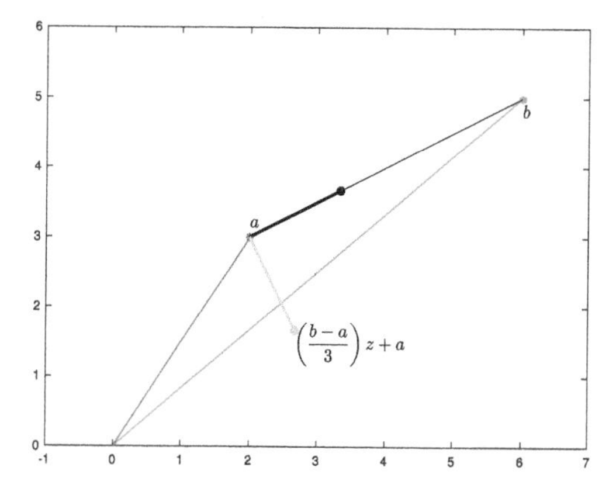

Figure 9.9 Rotating a scaled line segment.

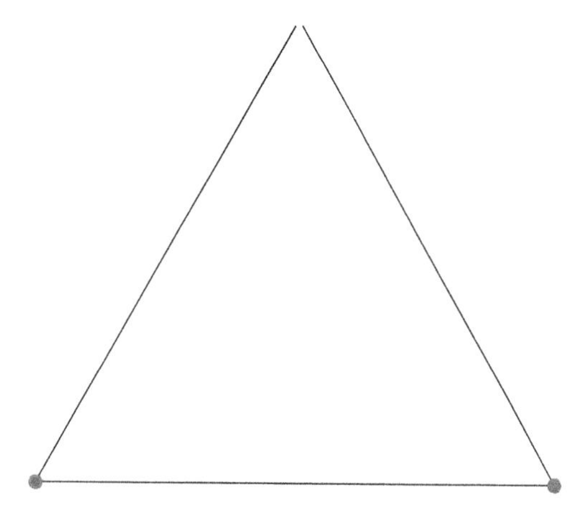

Figure 9.10 The classic Chaos Game board.

9.2. The Chaos Game

The Chaos Game is an interesting way to generate fractals. The classic Chaos Game is to use an equilateral triangle, and color each vertex a different color (e.g., red, yellow, and blue) (see Fig. 9.10). Color a standard, six-sided die so two faces are red, two are yellow, and two are blue.

Now choose a random point, called the **seed**. Classically it would be a point inside the triangle, but it does not matter, because after several turns, the points will be inside the triangle. Now roll the die. Move half the distance from the seed towards the vertex with the same color as what was rolled. Mark that point. Roll again, from the point you just marked, move half the distance towards the vertex with the same color as what was rolled. Mark that point. Repeat. (See Fig. 9.11.)

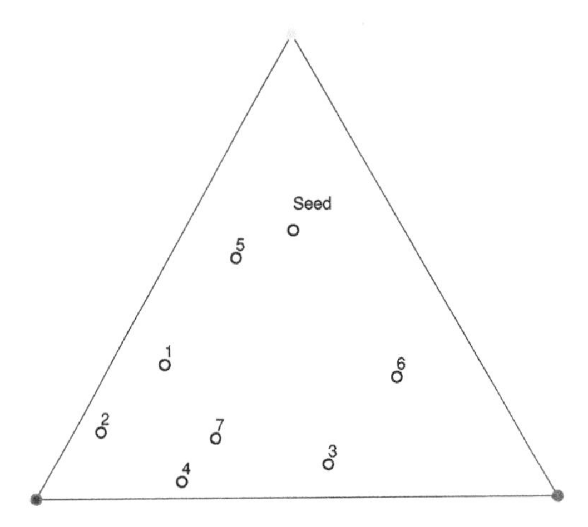

Figure 9.11 A few turns of the Chaos Game (with turns marked for demonstration).

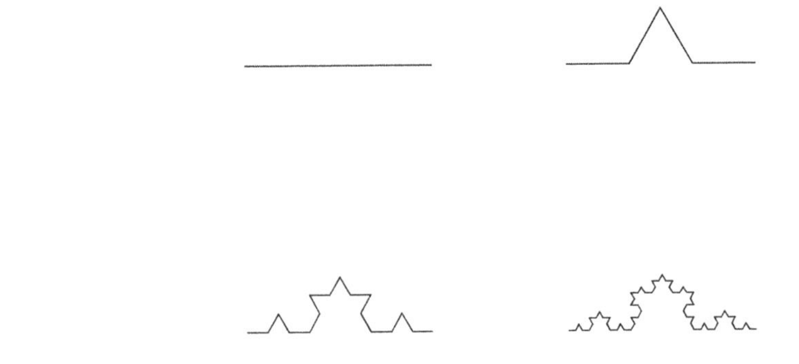

Figure 9.12 The Koch Curve for $n = 0, 1, 2, 3$.

After many iterations, the resulting picture may surprise you!

Modifications of this game is to use different polygons, and different distances towards the vertices to create different pictures.

9.3. Line replacement fractals

9.3.1 Snowflake fractals

The Koch Snowflake is a famous/familiar fractal. Several iterations of the Koch Curve are shown in Fig. 9.12. The snowflake is formed by repeating the Koch Curve around an equilateral triangle (see Fig. 9.13).

There are variations on this snowflake. One can have the base shape be a hexagon instead of a triangle (see Fig. 9.14).

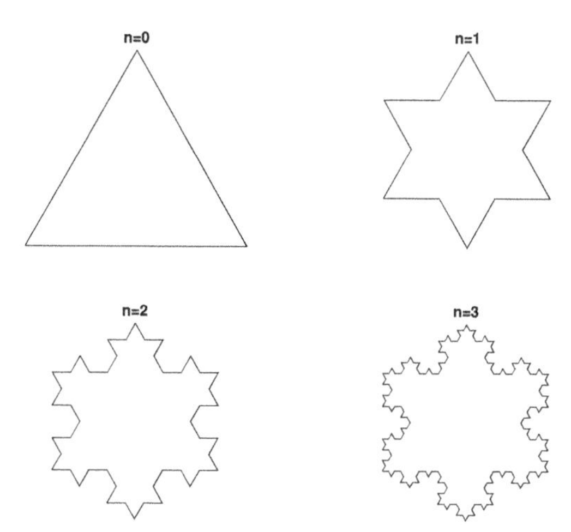

Figure 9.13 Several Iterations of the Koch Snowflake.

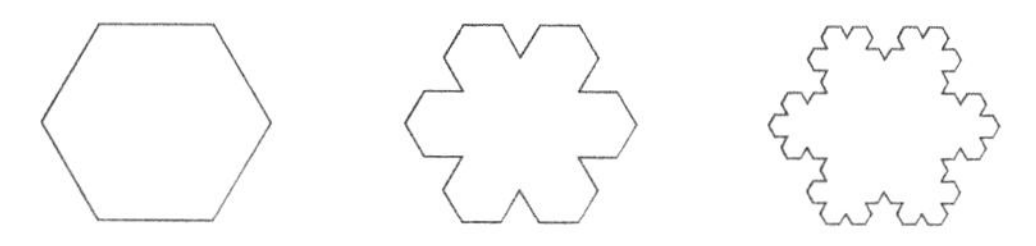

Figure 9.14 A variation of the snowflake.

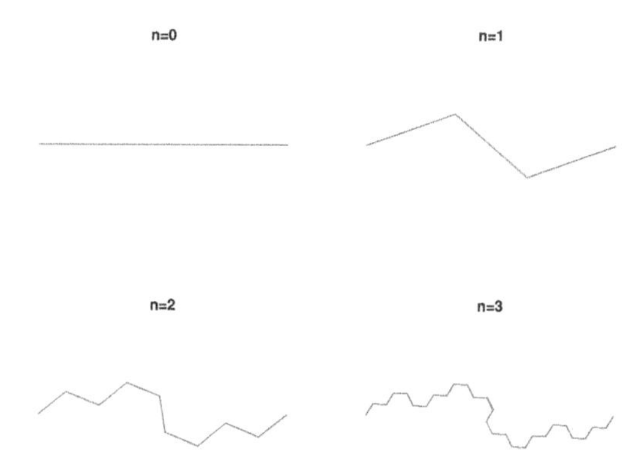

Figure 9.15 Several iterations of the Gosper Curve.

9.3.2 Gosper Island

Gosper Island is a fractal that can be considered another variation of the Koch Snowflake. It begins with a line segment and modifies the line segment at each iteration (see Figs. 9.15 and 9.16).

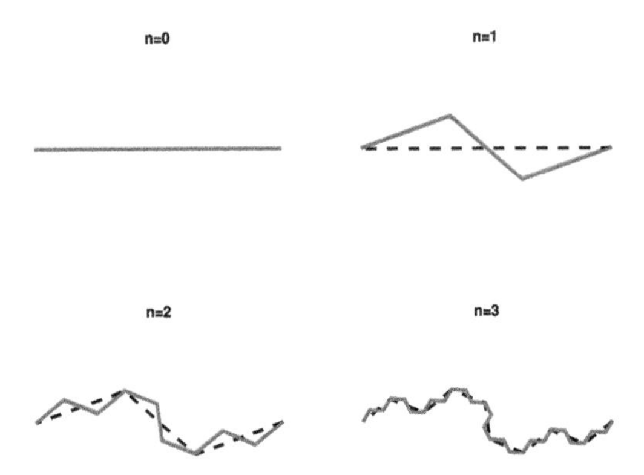

Figure 9.16 Another view of the Gosper Curve.

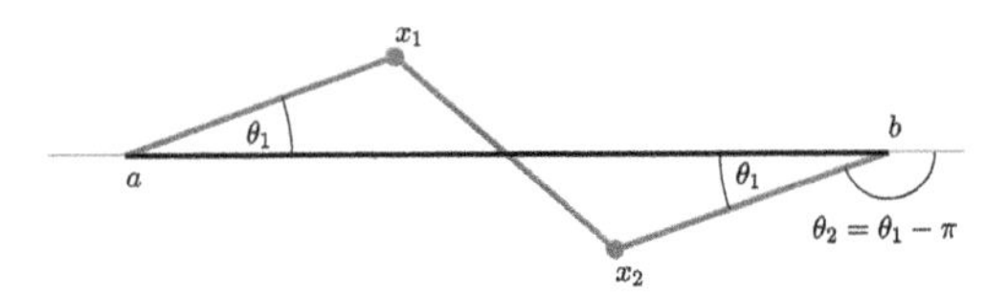

Figure 9.17 The Gosper Curve explained.

How one gets the line segments for the Gosper Curve is by taking the points (complex numbers) a and b and calculating the points x_1 and x_2 and connecting the points a, x_1, x_2, and b. The length of each subsequent line segment is the length of $\frac{b-a}{\sqrt{7}}$. The point x_1 is rotated from the segment $b-a$ by an angle of $\theta_1 = \tan^{-1}\left(\frac{\sqrt{3}}{5}\right)$. The point x_2 is rotated from the segment $b-a$ by an angle of $\theta_2 = \theta_1 - \pi$ (see Fig. 9.17). Then the Gosper Curve is repeated around a hexagon Figs. 9.18 and 9.19).

9.4. Geometric series

Geometric series are useful in many ways. Geometric series will help us to calculate things regarding snowflake fractals in the exercises.

Definition 9.4.1. A geometric series is a series of the form

$$\sum_{k=0}^{\infty} ar^k = a + ar + ar^2 + ar^3 + \cdots + ar^n + \cdots.$$

The following theorem is provided without proof and can be found in any calculus book.

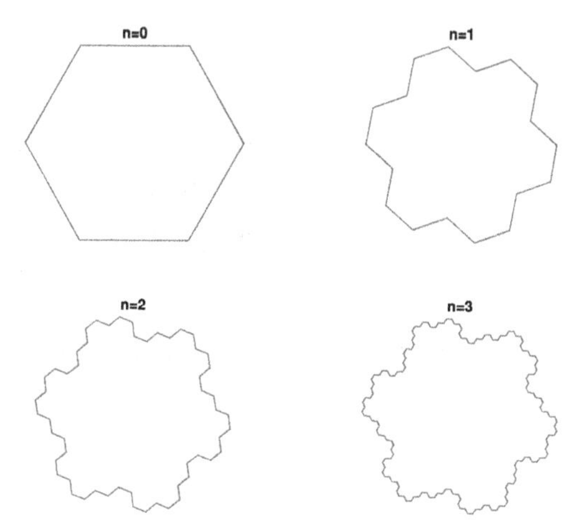

Figure 9.18 Several iterations of the Gosper Island.

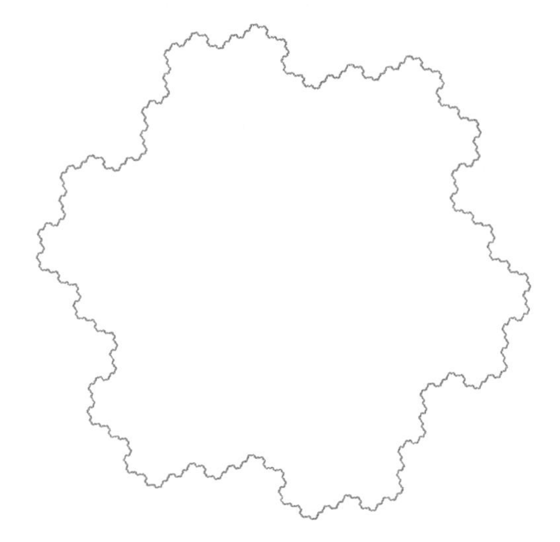

Figure 9.19 Gosper Island, $n = 6$.

Theorem 9.4.1. The geometric series

$$\sum_{k=0}^{\infty} ar^k = a + ar + ar^2 + ar^3 + \cdots + ar^n + \cdots$$

converges if and only if $|r| < 1$. In this case,

$$\sum_{k=0}^{\infty} ar^k = \frac{a}{1-r}.$$

9.5. Exercises

1. **Chaos Game** We will simulate the Chaos Game and use complex numbers. Our *seed* will be a complex number generated by the `rand` command for the real and imaginary values of the seed. You will create a function file called `chaosGame` that has one or more inputs. The first input is a positive integer n that will determine the number of "rolls of the die" there will be in the Chaos Game, or more inputs. The additional inputs, which are optional, will be marker specifications for drawing the points. The output of the function will be the time elapsed (generated by the commands `tic` and `toc`) to play the game. Your function file will do the following.

   (i) Capture the start time with `tic`.

   (ii) Perform the error check on the function input n and use the `error` command with appropriate message.

   (iii) Generate the three vertices of the equilateral triangle as $\cos\theta + i\sin\theta$ for θ from $\pi/2$ to $2\pi + \pi/2$, with an increment of $2\pi/3$. The triangle must be so the bottom base is horizontal. Draw the triangle in a solid black line.

   (iv) Store the default colors using the commands

   ```
   colors = lines(8); colors(8,:)=[0,0,0];
   ```

   (v) Draw the vertices using the "`o`" marker and color them using three of the default colors stored in `colors`. To make them more visible, specify the `MarkerFaceColor` and `MarkerEdgeColor` to be the appropriate color. You may want the `MarkerSize` to be larger than the default. (Hint: a `for` loop may be useful!)

   (vi) Preallocate a vector to the appropriate size to contain the "seed" and the n points generated by the n rolls of the die.

   (vii) Generate the seed to be in the first element of this vector by using the `rand` command. This seed should be a complex number. (How would we get a random complex number using the `rand` command?) Note that this seed point need not be inside the triangle.

   (viii) Now play the Chaos game for n turns: Use `randi` to simulate a roll of a standard die. The next element of your vector will be the point that is halfway between the current element and the vertex chosen by the "roll of the die". Note that you must make it so each vertex has an equal chance of being chosen by a roll of the die.

   (ix) After all of the turns are completed (thus the vector has been filled with the complex numbers obtained from the turns of the Chaos Game), either plot the points in the vector as black periods (`'k.'`) or use the `varargin` to plot the points.

   (x) The output of the function is the time elapsed as determined by `toc`.

(a) Play the game for $n = 10$ with no other input. Use `axis off` and `axis equal` commands. Put a meaningful title on the figure that has the number of iterations and time elapsed.

(b) Do the same as above but for $n = 200$, have the marker be "*" and set the `MarkerSize` to 3.

(c) Do the same as above but for $n = 30{,}000$ and specifying a black "o" with MarkerSize of 1.

2. **Chaos Game 2** We will simulate a modified version of the Chaos game and use complex numbers. We will use a square instead of a triangle and have eight "vertices" that are the corners and midpoints of the sides. Our *seed* will be a complex number generated by the `rand` command for the real and imaginary values of the seed. You will create a function file called `chaosGameSquare.m` that has one or more inputs. The first input is a positive integer n that will determine the number of "rolls of the die" there will be in the Chaos Game. The additional inputs, which are optional, will be marker specifications for drawing the points. The output of the function will be the time elapsed (generated by the commands `tic` and `toc`) to play the game. Your function file will do the following.

(i) Capture the start time with `tic`.

(ii) Perform the error check on the function input n and use the `error` command with appropriate message.

(iii) Store the vertices of the square in a vector: 1, $1+i$, i, $-1+i$, -1, $-1-i$, $-i$, $1-i$ (you may want to add extra 1 at the end to make it easier to draw...) Draw the square in a solid black line.

(iv) Store the default colors using the commands

```
colors = lines(8); colors(8,:)=[0,0,0];
```

(v) Draw the vertices using the "o" marker and color them using the default colors stored in `colors`. To make them more visible, specify the `MarkerFaceColor` and `MarkerEdgeColor` to be the appropriate color. You may want the `MarkerSize` to be larger than the default. (Hint: a `for` loop may be useful!)

(vi) Preallocate a vector to the appropriate size to contain the "seed" and the n points generated by the game.

(vii) Generate the seed to be in the first element of this vector by using the `rand` command. This seed should be a complex number. (How would we get a random complex number using the `rand` command?)

(viii) Generate n rolls of an *eight-sided die* in a vector `rolls`.

(ix) Now play the Chaos Game for n turns: Using the `rolls` vector to determine the chosen vertex, the next element of your vector of points will be gone from the previous element to $1/3$ of the distance between the previous element and the chosen vertex.

(x) After all of the turns are completed (thus the vector has been filled with the complex numbers obtained from the turns of the Chaos Game), either plot the points in the vector as periods (`'.'`) or use the `varargin` to plot the points.

(xi) The output of the function is the time elapsed as determined by `toc`.

Once the function is created, play the game as follows.

(a) Play the game for $n = 50$ with no other inputs. Use `axis off` and `axis equal` commands. Put a meaningful title on the figure that has the number of iterations and time elapsed (using `sprintf`).

(b) Do the same as above but for $n = 1000$.

(c) Do the same as above but for $n = 50{,}000$ and specifying a black period with MarkerSize of 2.

3. **Chaos Game 3** We will simulate a modified version of the Chaos game and use complex numbers. We will use a pentagon instead of a triangle. The function will be `chaosGamePent`. We will do as in Exercise 2 except for the initial shape and vertices, the die will be a *five-sided* die, and in step 2ix, the next element will be the point that is 3/8 the distance between the current element and the vertex chosen by the "roll of the die." Note that the definition of the pentagon and its vertices should be such that it lies flat on the bottom of the figure. This can be done with the equation $\sin\theta + i\cos\theta$ for $\theta = k(2\pi/5)$ for $k = 1, 2, \ldots, 5$.

   (a) Play the game for $n = 100$ with no other inputs. Use `axis off` and `axis equal` commands. Put a meaningful title on the figure that has the number of iterations and time elapsed (using `sprintf`).

   (b) Do the same as above but for $n = 1000$.

   (c) Do the same as above but for $n = 30{,}000$ and specifying a period with MarkerSize of 1.

4. **Chaos Game 4** We will simulate a modified version of the Chaos game and use complex numbers. We will use a hexagon instead of a triangle. The function will be `chaosGameHex`. We will do as in Exercise 2 except for the initial shape and vertices, the die will be a standard die, and in step 2ix, the next element will be the point that is 1/3 the distance between the current element and the vertex chosen by the "roll of the die." Note that the definition of the hexagon and its vertices should be such that it lies flat on the bottom of the figure. This can be done with the equation $\cos\theta + i\sin\theta$ for $\theta = k(\pi/3)$ for $k = 1, 2, \ldots, 6$.

   (a) Play the game for $n = 100$ with no other inputs. Use `axis off` and `axis equal` commands. Put a meaningful title on the figure that has the number of iterations and time elapsed (using `sprintf`).

   (b) Do the same as above but for $n = 1000$.

   (c) Do the same as above but for $n = 30{,}000$ and specifying a period with MarkerSize of 1.

5. **Koch Snowflake** We will create a function `kochSnowflake` that will have a **nested function** `kochpoints` that will draw the Koch Snowflake fractal at the nth iteration. The input for `kochSnowflake` will be a nonnegative integer n (return an appropriate error message if n is not a nonnegative integer). The output will be the time elapsed using the `tic` and `toc` commands.
   Pseudocode for function `kochSnowflake`
   - Capture the start time.
   - Check that n is a nonnegative integer; if not, give appropriate error message using the `error` command.
   - Create a vector v of vertices of the equilateral triangle; calculated by $\cos(k\theta) + i\sin(k\theta)$ for $k = 1, \ldots 4$ (so they connect) for the appropriate θ or by using your `mycircle2` function. The triangle must be so the bottom base is horizontal.
   - If $n = 0$, plot the triangle. Calculate the time elapsed and use the `return` command.
   - Initialize a vector S to be the first vertex of the triangle (stored in v) and create a loop that will use the **nested function** `kochpoints` on the last element of S, the next vertex of the triangle (stored in v) for n iterations. If we looped through using the nested function `kochpoints` on each of the vertices of the hexagon, there would be repeated points. Thus for each of the three sides of the triangle, we replace S with S minus the last element and concatenate it with the points from the nested function `kochpoints` using the last element of S and the next vertex (stored in v) using the given n.
   - Plot the sets of vertices S.
   - Capture the time elapsed for the output of the function.

   Pseudocode for **nested function** `kochpoints`: has inputs a, b, and n and output vector V.
   - Define the output V to have the elements a and b.
   - Repeat the following n times:
     - Define an empty vector B.
     - From 1 to [(number of elements in V) $-$ 1]
       * let $a =$ the current element of V and $b =$ the next element of V.
       * Calculate $x1$ to be the point that is $1/3$ of the way from a towards b.
       * Calculate $x2$ to be the point that is the midpoint between a and b MINUS $\sqrt{3}/2i(b-a)/3$.
       * Calculate $x3$ to be the point $2/3$ of the way from a towards b.
         (OR USE ANY DERIVATIONS FOR $x1$, $x2$, and $x3$ we did in class or you come up with on your own.) As long as it gives the proper picture, it is fine. Keep in mind that your derivation should have as few calculations as possible in order to speed up the code. To speed up the code even more, if you are using the same calculations in several of these formulas,

perform the calculation first, storing it as some variable, and then use the variable within the calculations.

* Concatenate B with the elements a, $x1$, $x2$, and $x3$.

– Now we redefine V. The new V will be B concatenated with the last element of the current V.

Once the function is created, we will visualize the snowflake as follows.

(a) Use `subplot` to have two rows, two columns of the snowflake plotted for $n = 0, 1, 2, 3$ (capture the time elapsed for each). Hint: it may be easiest to have a `for` loop for this.

- For each snowflake/subplot, capture the time elapsed.
- For each snowflake/subplot, use the `axis equal` and `axis off` commands.
- For each snowflake/subplot, create a title with "$n =$ __, y s" where y is the time elapsed to create that subplot. For example, you may see a title such as "$n = 3$, 0.3456 s" because the third iteration of the snowflake took 0.3456 s (made up numbers).

(b) Create another plot (NOT A SUBPLOT) that for $n = 5$ with a similar title as above that gives the number of iterations and time elapses, using `axis equal` and `axis off`.

(c) **Calculations** For the snowflake, calculate the following, where iteration $n = 0$ is the triangle. Show all work on paper (turned in).

i. Create a chart that has for each iteration n the number of sides/line segments, the number of new triangles that replaces part of the line segment at that iteration (this will be 0 for $n = 0$), the length of the side of the triangle at this stage ($\sqrt{3}$ for $n = 0$), and the area of the triangle. Do this for $n = 0$, $n = 1$, $n = 2$, $n = 3$, and $n = k$ for some $k \geq 1$.

ii. The area A_n enclosed by the snowflake at iteration n. Also find the limit $A = \lim_{n\to\infty} A_n$, if it exists.

iii. The perimeter P_n of the snowflake at iteration n. Also find the limit $P = \lim_{n\to\infty} P_n$, if it exists.

6. Fractal Snowflake We will create a function `snowflake` that will have a **nested function** `snowflakepoints` that will draw the Snowflake fractal at the nth iteration. The input for `snowflake` will be a nonnegative integer n (return an appropriate error message if n is not a nonnegative integer). Any additional inputs will be plot specifications. The output will be the time elapsed using the `tic` and `toc` commands.

Pseudocode for function `snowflake`:

- Capture the start time.
- Check that n is a nonnegative integer; if not, give appropriate error message using the `error` command.

- Create a vector v of vertices of a hexagon; calculated by $\cos(k\theta) + i\sin(k\theta)$ for $k = 1, \ldots, 7$ (so they connect) for the appropriate θ or by using your `mycircle2` function. The hexagon must be so the bottom base is horizontal.
- If $n = 0$, plot the hexagon. Calculate the time elapsed and use the `return` command to end the function.
- Initialize a vector S to be the first vertex of the hexagon (stored in v) and create a loop that will use the **nested function** `snowflakepoints` on the last element of S and the current vertex of the hexagon stored in v for n iterations. If we looped through using the nested function `snowflakepoints` on each of the vertices of the hexagon, there would be repeated points. Thus for each of the sides of the hexagon, we replace S with S minus the last element and concatenate it with the points from the nested function `snowflakepoints` using the last element of S and the next vertex (stored in v) using the given n.
- Plot the sets of vertices S, using `varargin` for the optional inputs if needed.
- Capture the time elapsed for the output of the function.

Pseudocode for **nested function** `snowflakepoints`: has inputs a, b, and n and output vector V.

- Define the output V to have the elements a and b.
- Repeat the following n times:
  - Define an empty vector B.
  - From 1 to [(number of elements in V) − 1]
    * Let $a =$ the current element of V and $b =$ the next element of V.
    * Calculate $x1$ to be the point that is 1/3 of the way from a towards b.
    * Calculate $x2$ to be the point that is the (midpoint between a and b) PLUS $\left(\frac{i\sqrt{3}}{2}\right)\left(\frac{(b-a)}{3}\right)$.
    * Calculate $x3$ to be the point 2/3 of the way from a towards b.
      (OR USE ANY DERIVATIONS FOR $x1$, $x2$, and $x3$ we did in class or you come up with on your own.) As long as it gives the proper picture, it is fine. Keep in mind that your derivation should have as few calculations as possible in order to speed up the code. To speed up the code even more, if you are using the same calculations in several of these formulas, perform the calculation first, storing it as some variable, and then use the variable within the calculations.
    * Concatenate B with the elements a, $x1$, $x2$, and $x3$.
  - Now we redefine V. The new V will be B concatenated with the last element of the current V.

Once the function is created, we will visualize the snowflake as follows.

(a) Use `subplot` to have 2 rows, 2 columns of the snowflake plotted for $n = 0, 1, 2, 3$ (capture the time elapsed for each). Hint: it may be easiest to have a `for` loop for this.

- For each snowflake/subplot, capture the time elapsed.
- For each snowflake/subplot, use the `axis equal` and `axis off` commands.
- For each snowflake/subplot, create a title with "$n =$ __, y s" where y is the time elapsed to create that subplot. For example, you may see a title such as "$n = 3$, 0.3456 s" because the third iteration of the snowflake took 0.3456 seconds (made up numbers).

(b) Create another plot (NOT A SUBPLOT) that for $n = 5$ and optional color "Deep Sky Blue" that has rgb(0, 191, 255) (remember how you need to convert to a vector for MATLAB to recognize the RGB color). Have a similar title as above that gives the number of iterations and time elapses, and also using `axis equal` and `axis off`. Set the background color to white using `set(gcf,'Color','w')`.

(c) EXTRA CREDIT: figure out how to make an animated gif of the iterations looping from $n = 0$ to $n = 5$ (no title).

7. Gosper Island We will create a function `gosper` that will have a nested function `gosperpoints` that will draw the Gosper Island fractal at the nth iteration. The input for `gosper` will be a nonnegative integer n (return an appropriate error message if n is not a nonnegative integer). The output will be the time elapsed using the `tic` and `toc` commands.

Pseudocode for `gosper`.

- Capture start time.
- Check that n is a nonnegative integer; if not, give appropriate error message.
- Create a vector v of vertices of a hexagon; calculated by $\cos(k\theta) + i\sin(k\theta)$ for $k = 1, \ldots 7$ (so they connect) for the appropriate θ or by using your `mycircle2` function.
- Initialize a vector G to be an empty vector.
- For each of the six sides of the hexagon (sides between points $v(k)$ and $v(k+1)$, use the nested function `gosperpointsLO`, to get the new vertices.
- Plot the sets of vertices G.
- Capture the time elapsed.

Pseudocode for `gosperpoints`: has inputs a, b, and n and output V.

- Let $\theta_1 = \tan^{-1}(\sqrt{3}/5)$, and $\theta_2 = \theta_1 - \pi$.
- Calculate the two rotations $r_1 = \cos(\theta_1) + i\sin(\theta_1)$, and $r_2 = \cos(\theta_2) + i\sin(\theta_2)$.
- Define the output V to have the elements a and b.
- Repeat the following n times:
  - Define an empty vector B;

- From 1 to (number of elements in $V - 1$)
  - let $a =$ the current element of V and $b =$ the next element of V.
  - Calculate $x = \frac{b-a}{\sqrt{7}}$.
  - Concatenate B with the elements a, $r_1 x + a$, $r_2 x + b$.
- Redefine V to be B concatenated with the last element of the current V.

Once the function is created, we will visualize Gosper Island as follows.

(a) Use `subplot` to have two rows, three columns of the Gosper Island plotted for $n = 0, 1, 2, 3, 4$, and 5 (capture the time elapsed for each).

(b) For each subplot, use the `axis equal` and `axis off` commands.

(c) For each subplot, create a title with "$n = x$, y s" where x is the value of n for that subplot, and y is the time elapsed to create that subplot.

(d) **Calculations** For the Gosper Island, calculate the following (stage $n = 0$ is the hexagon). Show all work on paper (turned in).

i. The number of points p_n that make up one curve at each state n (do not count repeats that may be in your code).

ii. The island is made up of six curves connected. Find the number of points x_n that make up the entire island at stage n (do not count repeats that may be in your code).

iii. The area A_n inside the island at stage n. Also find the limit $A = \lim_{n\to\infty} A_n$, if it exists.

iv. The perimeter P_n of the island at stage n. Also find the limit $P = \lim_{n\to\infty} P_n$, if it exists.

CHAPTER 10

Series and Taylor Polynomials

10.1. Review of series

The material will be more understandable after a review of sequences and series from calculus, for example Chapter 11 from [24]. Recall from Section 9.4 that a geometric series is a series of the form

$$\sum_{n=0}^{\infty} r^n = \sum_{n=1}^{\infty} r^{n-1}$$

and that a geometric series of this form converges to $\dfrac{1}{1-r}$ when $|r| < 1$ and is divergent otherwise.

From now on, the subscripts and superscripts on the summation notation will be left off. In other words, the notation $\sum a_n$ will stand for $\sum_{n=0}^{\infty} a_n$.

Another useful type of series is the alternating series.

Definition 10.1.1. An **alternating series** is a series of the form

$$\sum(-1)^n b_n \quad \text{or} \quad \sum(-1)^{n+1} b_n$$

where $b_n \geq 0$ for all n.

Example 10.1.1. A famous example of an alternating series is the alternating harmonic series:

$$\sum_{n=1}^{\infty}(-1)^{n+1}\frac{1}{n} = 1 - \frac{1}{2} + \frac{1}{3} - \frac{1}{4} + \cdots$$

Theorem 10.1.1 (Alternating series estimation theorem). Consider an alternating series $\sum(-1)^n b_n$ that satisfies

(a) $0 \leq b_{n+1} \leq b_n$ for all n, and

(b) $\lim_{n\to\infty} b_n = 0$

Then the alternating series converges. In addition, if we say

$s = \sum(-1)^n b_n$ and $s_n = \sum_{k=0}^{n}(-1)^k b_k$, then the remainder $R_n = s - s_n$

is estimated by

$$|R_n| = |s - s_n| \leq b_{n+1}.$$

Programming Mathematics Using MATLAB®
https://doi.org/10.1016/B978-0-12-817799-0.00016-8

The proof of this is not given here. Notice that many times (and indeed, even in lecture), this is split up into two theorems; the first with the statement giving the sufficient conditions for an alternating series to converge, and the second has the statement about the remainder. For brevity's sake it is combined here into one theorem.

Example 10.1.2. The alternating harmonic series converges by the alternating series estimation theorem, as $b_n = \frac{1}{n}$ is a decreasing sequence that converges to 0. In addition, if we looked at

$$s_5 = 1 - \frac{1}{2} + \frac{1}{3} - \frac{1}{4} + \frac{1}{5} = \frac{47}{60} = 0.78\bar{3},$$

then $|R_5| \leq b_6 = \frac{1}{6}$. Thus the actual sum s of the series is somewhere in the interval

$$\frac{47}{60} \pm \frac{1}{6}, \text{ or between } \frac{37}{60} \text{ and } \frac{57}{60}, \text{ or between } 0.61\bar{6} \text{ and } 0.95.$$

Recall that a series $\sum a_n$ is **absolutely convergent** if $\sum |a_n|$ converges. An important theorem tells us that any absolutely convergent series converges, but not necessarily vice versa. This gives us the term **conditionally convergent**; a series is conditionally convergent if $\sum a_n$ converges but $\sum |a_n|$ diverges.

Example 10.1.3. The alternating harmonic series is conditionally convergent.

Two useful tests for convergence are the ratio and root tests. There are, of course, others but for now we will remind you of these two without proof.

Theorem 10.1.2 (Ratio test). Consider the series $\sum a_n$ and let

$$\lim_{n\to\infty} \left| \frac{a_{n+1}}{a_n} \right| = L.$$

(a) If $L < 1$, the series $\sum a_n$ is absolutely convergent.
(b) If $L > 1$ (including $L = \infty$), the series $\sum a_n$ diverges.
(c) If $L = 1$, the ratio test is not applicable.

Theorem 10.1.3 (Root test). Consider the series $\sum a_n$ and let

$$\lim_{n\to\infty} \sqrt[n]{|a_n|} = L.$$

(a) If $L < 1$, the series $\sum a_n$ is absolutely convergent.
(b) If $L > 1$ (including $L = \infty$), the series $\sum a_n$ diverges.
(c) If $L = 1$, the ratio test is not applicable.

Example 10.1.4. Consider the series $\sum_{n=0}^{\infty} \frac{(-1)^n}{n!}$.

We have

$$\begin{aligned}\lim_{n\to\infty}\left|\frac{a_{n+1}}{a_n}\right| &= \lim_{n\to\infty}\frac{\frac{1}{(n+1)!}}{\frac{1}{n!}}\\ &= \lim_{n\to\infty}\frac{n!}{(n+1)!}\\ &= \lim_{n\to\infty}\frac{1}{n+1}=0.\end{aligned}$$

Since this limit equals zero, which is certainly less than one, the series is absolutely convergent by the ratio test.

All of the above are useful, but they are really used in the context of *power series*, and in particular, *Taylor series*.

10.2. Power series

Definition 10.2.1. Recall the definition of a **power series centered at** $x_0 \in \mathbb{R}$:

$$\sum_{n=0}^{\infty} a_n(x-x_0)^n = a_0 + a_1(x-x_0) + a_2(x-x_0)^2 + a_3(x-x_0)^3 + \cdots$$

The real numbers $a_0, a_1, a_2, \ldots$ are called the **coefficients of the power series.**

The main questions about power series are: "For what values of x does the series converge?" and "What does the series equal?" Knowing these answers allow us to see why we care about power series. In other words, when does the infinite sum make sense, and given a value of x, when does it equal some real number, even if it is difficult to find that exact real number? Notice that every power series converges at $x = x_0$. But if that were the only case, then power series would not be useful at all. So we want power series that converge for additional values of x.

The goal is to REPRESENT A FUNCTION, ESPECIALLY A COMPLICATED FUNCTION, AS A POWER SERIES. This is tremendously useful for approximations and calculations and many other results beyond the scope of this course.

Theorem 10.2.1. Consider the power series $\sum_{n=0}^{\infty} a_n(x-x_0)^n$.

(a) If the power series converges at $c \neq x_0$, then it converges for all x such that $|x-x_0| < |c-x_0|$.

(b) If the power series diverges at $d \in \mathbb{R}$, it diverges for all x such that $|x-x_0| > |d-x_0|$.

Proof. Suppose there exists $c \neq x_0$ such that the power series converges. We will show that, for every x such that $|x-x_0| < |c-x_0|$, the series converges.

Let $x \in \mathbb{R}$ such that $|x-x_0| \leq r < |c-x_0|$. We know that $\sum_0^\infty a_n(c-x_0)^n$ is convergent, so

$$\lim_{n\to\infty} a_n(c-x_0)^n = 0.$$

In particular, there is a constant M such that $|a_n(c-x_0)^n| \leq M$ for all n. Thus

$$|a_n|\,|x-x_0|^n = |a_n|\,|x-x_0|^n \left(\frac{|x-x_0|}{|c-x_0|}\right)^n \leq M\left(\frac{r}{|c-x_0|}\right)^n.$$

Since $\frac{r}{|c-x_0|} < 1$, we have then that $\sum M\left(\frac{r}{|c-x_0|}\right)^n$ is a convergent geometric series, so we have $\sum a_n(x-x_0)^n$ is absolutely convergent.

Now suppose that there exist $d \in \mathbb{R}$ such that the power series diverges. Suppose there exists $w \in \mathbb{R}$ with $|d-x_0| < |w-x_0|$ and the power series converges at $x = w$. By above, the series must converge at d, which is a contradiction. Thus there cannot exist any such w. □

The above theorem is useful, but is mostly used to prove the following theorem.

Theorem 10.2.2. For a given power series $\sum_{n=0}^{\infty} a_n(x-x_0)^n$, there are only three possibilities:

(a) The series converges only when $x = x_0$.

(b) The series converges for all $x \in \mathbb{R}$.

(c) There exist $R > 0$ such that the series converges if $|x-a| < R$ and diverges if $|x-a| > R$.

Proof. Consider Theorem 10.2.1. If no such $c \neq x_0$ exists, we have case 1 of the theorem. If no such d exists, we have case 2 of the theorem.

Now suppose we have both a point $c \neq x_0$ where the series converges and a point d where the series diverges. Let $C = |c-x_0|$ and $D = |d-x_0|$ and consider the intervals $I_C = (x_0 - C, x_0 + C)$ and $I_D = [x_0 - D, x_0 + D]$. (Draw a picture!)

Let $S = \{x \in \mathbb{R} \mid \text{series converges}\}$. This set is non-empty since we have all $x \in S$ for all $x \in I_C$ by Theorem 10.2.1. By the same theorem, $x \notin S$ for all $x \notin I_D$. Thus $x_0 + D$ is an upper bound for S since for all $x > x_0 + D$, the series diverges so is not in S. By the completeness axiom of real numbers, S has a "least upper bound", which we will call R. We know that $R > 0$. This tells us if $|x - x_0| < R$, $x \in S$, and if $|x - x_0| > R$, $x \notin S$, and the result is proven. □

In the third case, what happens when $|x - a| = R$ depends on the series and you have to check. In other words, some series converge at the endpoints, others diverge, and others converge at one endpoint and diverge at the other.

The number R is called the **radius of convergence** and in the first case of Theorem 10.2.2, we say $R = 0$. In the second case of Theorem 10.2.2 we say $R = \infty$. In the third case, you have four possibilities for the **interval of convergence**:

$$(x_0 - R, x_0 + R), \qquad [x_0 - R, x_0 + R], \qquad (x_0 - R, x_0 + R], \qquad [x_0 - R, x_0 + R).$$

This theorem is important because it tells us that a power series will never only converge at a few points, or only on the integers, etc.

Example 10.2.1. Consider the power series $\sum \frac{(-1)^n x^n}{n!}$. We see that

$$\begin{aligned}
\lim_{n\to\infty}\left|\frac{a_{n+1}}{a_n}\right| &= \lim_{n\to\infty}\frac{\frac{|x|^{n+1}}{(n+1)!}}{\frac{|x|^n}{n!}} \\
&= \lim_{n\to\infty}\frac{n!|x|}{(n+1)!} \\
&= \lim_{n\to\infty}\frac{|x|}{n+1} = 0.
\end{aligned}$$

Thus the series is absolutely convergent for any $x \in \mathbb{R}$ by the ratio test. Thus the radius of convergence is $R = \infty$ and the interval of convergence is $I = (-\infty, \infty)$.

Example 10.2.2. Compare the radii and intervals of convergence of the following series. The work is shown for one of the series. Can you see how to get the answers for the others?

(a) $\sum_{n=0}^{\infty}(x-7)^n$. $R = 1$, $I = (6, 8)$.

(b) $\sum_{n=1}^{\infty}\frac{(x-7)^n}{n}$, $R = 1$, $I = [6, 8)$.

$$\lim_{n\to\infty}\left|\frac{a_{n+1}}{a_n}\right| = \lim_{n\to\infty}\frac{\frac{|x-7|^{n+1}}{n+1}}{\frac{|x-7|^n}{n}}$$

$$= \lim_{n\to\infty} \frac{n|x-7|}{n+1} = |x-7|.$$

By the ratio test, the power series converges when $|x-7| < 1$, and diverges when $|x-7| > 1$ and so $R = 1$. But when $|x-7| = 1$, i.e., when $x = 6$ and $x = 8$, the ratio test is not applicable so we look at each of these cases separately. When $x = 6$, the power series $\sum \frac{(x-7)^n}{n} = \sum \frac{(-1)^n}{n}$ which is the alternating harmonic series, which converges. When $x = 8$, the power series is the harmonic series, which diverges. Thus $I = [6, 8)$.

(c) $\sum_{n=1}^{\infty} \frac{(x-7)^n}{n^2}$, $R = 1$, $I = [6, 8]$.

(d) $\sum_{n=0}^{\infty} n!(x-7)^n$, $R = 0$, $I = \{7\}$.

(e) $\sum_{n=0}^{\infty} \frac{(x-7)^n}{n!}$, $R = \infty$, $I = (-\infty, \infty)$.

Definition 10.2.2. A function f is **analytic** at $x = a$ if there exists a power series expansion centered at a with radius of convergence $R > 0$ that converges to $f(x)$ for $x \in (a-R, a+R)$.

The next theorem is very useful because it allows us to get series for more functions, and possibly use Taylor polynomials (see Section 10.3 below) for approximating these functions.

Theorem 10.2.3. If $f(x)$ is analytic (can be represented as a power series) with $f(x) = \sum a_n(x-a)^n$ and $R > 0$, then $f'(x)$ and $\int f(x)\,dx$ can also be represented as a power series with radius of convergence also R. In addition, for all $x \in (a-R, a+R)$,

$$f'(x) = \sum_{n=1}^{\infty} na_n(x-a)^{n-1} = a_1 + 2a_2(x-a) + 3a_3(x-a)^2 + \cdots$$

and

$$\int f(x)\,dx = C + \sum_{n=0}^{\infty} a_n \frac{(x-a)^{n+1}}{n+1} = C + a_0(x-a) + a_1 \frac{(x-a)^2}{2} + \cdots$$

Note that this theorem says

$$\frac{d}{dx}\left(\sum a_n(x-a)^n\right) = \sum \frac{d}{dx} a_n(x-a)^n$$

and

$$\int \left(\sum a_n(x-a)^n\right) dx = \sum \int a_n(x-a)^n dx.$$

IMPORTANT: the theorem tells us the radius of convergence stays the same, yet the actual interval of convergence may change as demonstrated in the following example.

Example 10.2.3. Consider the series $\sum_{n=1}^{\infty} \frac{(x-7)^n}{n}$ from Example 10.2.2 above. We got $R=1$, $I=[6,8)$. Let $f(x)=\sum_{n=1}^{\infty} \frac{(x-7)^n}{n}$. Theorem 10.2.3 tells us that, for all $x \in (6,8)$, the derivative exists and is

$$f'(x)=\sum_{n=1}^{\infty} \frac{n(x-7)^{n-1}}{n} = \sum_{n=1}^{\infty} (x-7)^{n-1} = \sum_{n=0}^{\infty} (x-7)^n.$$

Thus $R=1$ for the series for $f'(x)$. Upon further investigation, we see that the interval of convergence for this series for $f'(x)$ is now $(6,8)$, which is different from the interval of convergence for $f(x)$.

10.3. Taylor polynomials and Taylor series

The rest of the theorems in this chapter will not be proven. Proofs can be found in calculus and/or analysis texts.

Definition 10.3.1. For a given function $f(x)$ and $a \in \mathbb{R}$, the **Taylor series of $f(x)$ about $x=a$** is defined as:

$$\sum_{k=0}^{\infty} \frac{f^{(k)}(a)}{k!}(x-a)^k = f(a) + f'(a)(x-a) + \frac{f''(a)}{2!}(x-a)^2 + \cdots$$

Here we are assuming that f is infinitely differentiable at a. We are also using the convention $0! = 1$, and $f^{(0)}(x) = f(x)$.

Taylor series are useful in their own right, but are most useful when looking at the partial sums of the series, known as Taylor polynomials.

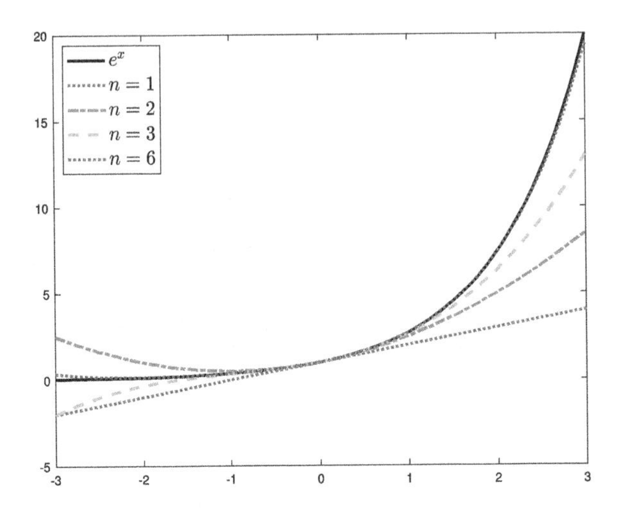

Figure 10.1 Taylor polynomials for $y = e^x$.

Definition 10.3.2. For a given function $f(x)$ and $a \in \mathbb{R}$, the **Taylor polynomial of degree n about** $x = a$ is defined as

$$T_n(x) = \sum_{k=0}^{n} \frac{f^{(k)}(a)}{k!}(x-a)^k$$
$$= f(a) + f'(a)(x-a) + \frac{f''(a)}{2!}(x-a)^2 + \cdots + \frac{f^{(n)}(a)}{n!}(x-a)^n.$$

Here we are assuming that f is differentiable n times at a. Then we let $R_n(x) = f(x) - T_n(x)$ be the **remainder of the Taylor polynomial**. Taylor series with center $a = 0$ are commonly called **Maclaurin series**.

Example 10.3.1. The Maclaurin series for $f(x) = e^x$ becomes

$$e^x = \sum \frac{x^n}{n!}$$

since $f^{(n)}(0) = 1$ for all n. Fig. 10.1 above shows how Taylor polynomials are decent approximations for the exponential function near $x = 0$, and the higher degree of Taylor polynomial used, the better this approximation becomes, even for values further away from $x = 0$.

The next theorem relates general power series to Taylor series.

Theorem 10.3.1 (Taylor series theorem)**.** Suppose that a power series $\sum a_n(x-a)^n$ has a radius of convergence $0 < R \leq \infty$ $(R > 0)$. Then, the function $f(x)$ that this series represents is infinitely differentiable on $(a-R, a+R)$ and its coefficients a_n are given by

$$a_n = \frac{f^{(n)}(a)}{n!}, \qquad n = 0, 1, 2, \ldots$$

Theorem 10.3.2. Using the notation from above, if

$$f(x) = T_n(x) + R_n(x), \text{ and } \lim_{n\to\infty} R_n(x) = 0$$

for $|x-a| < R$, then f is **equal** to the sum of the Taylor series:

$$\begin{aligned} f(x) &= \sum_{k=0}^{\infty} \frac{f^{(k)}(a)}{k!}(x-a)^k \\ &= f(a) + f'(a)(x-a) + \frac{f''(a)}{2!}(x-a)^2 + \cdots + \frac{f^{(n)}(a)}{n!}(x-a)^n + \cdots. \end{aligned}$$

The key is to get an estimate of $R_n(x)$ in order for these series to be useful.

Theorem 10.3.3 (Taylor's theorem)**.** Suppose that, for any $n \in \mathbb{N}$, $f(x)$ has $n+1$ derivatives in a neighborhood of $x = a$. If

$$\begin{aligned} T_n(x) &= \sum_{k=0}^{n} \frac{f^{(k)}(a)}{k!}(x-a)^k \\ &= f(a) + f'(a)(x-a) + \frac{f''(a)}{2!}(x-a)^2 + \cdots + \frac{f^{(n)}(a)}{n!}(x-a)^n \end{aligned}$$

then $f(x) = T_n(x) + R_n(x)$, where

$$R_n(x) = \frac{f^{(n+1)}(c)}{(n+1)!}(x-a)^{n+1}$$

for some c between x and a. If $x = a$, then $c = a$.

Note: this formula for the remainder is known as Lagrange's form of the remainder. There are other forms of the remainder. Here are two other famous forms:

$$R_n(x) = \frac{f^{(n+1)}(c)}{n!}(x-c)^n(x-a) \qquad \text{Cauchy's form,}$$

$$R_n(x) = \frac{1}{n!}\int_a^x f^{(n+1)}(t)(x-t)^n\,dt \qquad \text{Integral form.}$$

Here is another way one can estimate the error when using Taylor polynomials to estimate the function represented by a Taylor series:

Theorem 10.3.4 (Taylor's inequality). Suppose $f(x)$ is represented as a power series centered at a with $R > 0$. If $|f^{(n+1)}(x)| \leq M$ for all x with $|x - a| \leq \delta$, then the remainder $R_n(x)$ of the Taylor series satisfies

$$|R_n(x)| \leq \frac{M}{(n+1)!}|x - a|^{n+1} \quad \text{for } |x - a| \leq \delta.$$

If you read the above theorems carefully, you may notice that it is NOT quite the case that "f is analytic iff f is infinitely differentiable. Here is a counterexample:

Example 10.3.2. Center the series at $a = 0$ for $f(x)$:

$$f(x) = \begin{cases} e^{-1/x} & x > 0, \\ 0 & x \leq 0. \end{cases}$$

It can be shown (can you show it?) that, while f is infinitely differentiable, even at 0, there is no power series (with $R > 0$) centered at 0 (look at Lagrange's form of the remainder for $x > 0$ close to 0).

The good news is that many functions that we are familiar with CAN be represented by a Taylor or Maclaurin series.

Common Maclaurin series (look up others in your calculus book):

$$e^x = \sum_{n=0}^{\infty} \frac{x^n}{n!} \qquad R = \infty,$$

$$\cos x = \sum_{n=0}^{\infty} \frac{(-1)^n x^{2n}}{(2n)!} \qquad R = \infty,$$

$$\sin x = \sum_{n=0}^{\infty} \frac{(-1)^n x^{2n+1}}{(2n+1)!} \qquad R = \infty,$$

$$\tan^{-1} x = \sum \frac{(-1)^n x^{2n+1}}{2n+1} \qquad R = 1,$$

$$\frac{1}{1-x} = \sum_{n=0}^{\infty} x^n \qquad R = 1.$$

One can use these series to get an estimate for e:

$$e \approx \sum_{k=0}^{n} \frac{1}{k!}.$$

The following examples show the usefulness of Theorem 10.2.3 when applied to Maclaurin and Taylor series.

Example 10.3.3. Use Maclaurin series to prove the following statements.

(a) Prove $\frac{d}{dx}e^x = e^x$.

(b) Prove that $\cos x$ is an even function and that $\frac{d}{dx}\cos x = -\sin x$.

(c) Prove that $\sin x$ is an odd function and that $\frac{d}{dx}\sin x = \cos x$.

Example 10.3.4. What is a Maclaurin series for $\ln(1-x)$ centered at 0? What is its radius of convergence?

Example 10.3.5. Consider the geometric series $\sum_{n=0}^{\infty} x^n = \frac{1}{1-x}$, for $|x| < 1$.

(a) Use the geometric series above to find a series for $\frac{1}{1+x^2}$. What is the radius of convergence?

(b) Use Theorem 10.2.3 to find a series for $\tan^{-1} x$. What is the radius of convergence? Interval of convergence?

(c) Prove the Leibniz series identity: $\sum_{k=0}^{\infty} \frac{(-1)^k}{2k+1} = \frac{\pi}{4}$.

Example 10.3.6. Using the Maclaurin series for $\ln(1+x)$, show that

$$\ln 2 = \sum_{k=1}^{\infty} \frac{(-1)^{k+1}}{k}.$$

10.4. Exercises

1. Create a function `isnnInt` that will have one input, n, and the output will be a logical true/false as to whether n is a nonnegative integer (but not necessarily of data type integer; data type double is still allowed). Note that no error messages are displayed with this function; it returns either `true` or `false`.
2. This problem looks at the Leibniz series

$$\frac{\pi}{4} = \sum_{k=0}^{\infty} \frac{(-1)^k}{2k+1}.$$

The nth partial sum of the series can be used to estimate $\pi/4$ and thus other values with π (this is a crude estimation; other/better estimations are a topic for another course). Notice that it is an alternating series. Your `leibniz` function will have the following features.

- The function has one input, n, which should be a **nonnegative integer**. Use your function `isnnInt` to check it; if not, an appropriate error message is displayed using the `error` command.
- The function will have **two** outputs. The first output is the nth partial sum of the Leibniz series:

$$\sum_{k=0}^{n} \frac{(-1)^k}{(2k+1)} \approx \frac{\pi}{4}.$$

  A loop can be used for this calculation but it may be fastest if you use vectorized code instead. The second output will be the estimated error of this partial sum using the alternating series estimation theorem.

3. The next function `nleibniz2` will not calculate partial sums, but instead tell you what n should equal used in order to use the `leibniz` function to approximate $\pi/4$ to a certain degree of accuracy ("tolerance level").
   - The `nleibniz2` function has *one* input: a "tolerance level" ε for using the partial sum.
   - Your function should check that $\varepsilon > 0$; if not, an appropriate error message is displayed using the `error` command.
   - The function figures out the minimum value of n (obtained from the alternating series estimation theorem) for the nth partial sum to have an error less than or equal to ε. The output of the function is this value n.
4. Use your function `leibniz` to estimate π and its error. (Careful! You need to adjust your answers from the answer you get from your function for this!). Note that even though MATLAB® has an accurate estimation of π, do not use this to estimate the error. Use $n = 3$, $n = 9$, $n = 99$, and $n = 9999$. Use the `fprintf` command to make your answers appear like below, with the same number of decimal places shown (it may be more convenient to have more than one `fprintf` command). Note that the numbers below are made up for display purposes.

```
Using the partial sum of the Leibniz series:
 n    Estimate      Error estimate   Between
 3    3.123456789  1.123456789      2.00000000 and 4.24691358
 9    3.143456789  0.123456789      3.02000000 and 3.26691358
 99   3.141556789  0.012345678      3.12921111 and 3.15390247
 9999 3.141596789  0.001234567      3.14036222 and 3.14283136
```

5. Using `nleibniz2`, what value of n should be used to get the estimate of π accurate to six decimal places? Careful! What would the error tolerance need to equal to be within that many decimal places, and how do you adjust between error for $\pi/4$ and error for π?

6. This problem looks at the alternating harmonic series

$$\sum_{k=1}^{\infty} \frac{(-1)^{k+1}}{k} = \ln(2).$$

In calculus or analysis, using Taylor (Maclaurin) series on $f(x) = \ln(1+x)$ it can be shown that the Taylor series converges when $x = 1$ and you get the above equation. Therefore, the nth partial sum of the series can be used to estimate $\ln(2)$. Your `ln2` function will do the following:

   - The function will have one input n which should be a positive integer. Check for it using `isnnInt`; if not, an appropriate error message is displayed using the `error` command.
   - The function will have **two** outputs. The first output is the nth partial sum of the series:

$$\sum_{k=1}^{n} \frac{(-1)^{k+1}}{k}.$$

   A loop can be used for this calculation but it may be fastest if you use vectorized code instead. The second output will be the estimated error of this partial sum using the alternating series estimation theorem.

7. The next function `nln2` will not calculate partial sums, but instead tell you the minimum n should equal in order to use the `ln2` function to approximate $\ln(2)$ to a certain degree of accuracy ("tolerance level").
   - The `nln2` function has *one* input: a "tolerance level" ε for using the partial sum.
   - Your function should check that $\varepsilon > 0$; if not, an appropriate error message is displayed using the `error` command.
   - The function figures out the minimum value of n (determined by the alternating series estimation theorem) for the nth partial sum to have an error less than or equal to ε. The output of the function is this value n.
8. Use your functions `nln2` and `ln2` to estimate $\ln(8)$ accurate to four decimal places. Careful! What would the error tolerance need to equal to be within that many decimal places, and how do you adjust the error for $\ln(2)$ to get the estimate for $\ln(8)$? Use the value of n you get from `nln2` in your function `ln2` to estimate $\ln(8)$.
9. This problem is looking at Taylor/Maclaurin series and Taylor polynomials for $f(x) = \cos(x)$.

$$\cos(x) = \sum_{k=0}^{\infty} \frac{(-1)^k \, x^{2k}}{(2k)!}$$

Create a function called `mycos.m` that calculates the nth-degree Taylor polynomial for $f(x) = \cos(x)$ using the above Maclaurin series for $\cos(x)$. The function should have two inputs; x and n. Make sure that n is a **nonnegative integer** using your function `isnnInt`; if not, an appropriate error message is displayed using the `error` command. The calculations should be such that x can be a number, vector, or matrix. The input n is the **highest degree term that will appear in the polynomial**. Thus if $n=8$, the degree of the polynomial will be no bigger than 8. Notice that if we enter $n=9$, we should get the same as $n=8$. Use the `warning` command if an odd value of n is entered. Careful: if we want $n=8$, what should k equal for the last term in the sum? If we input $n=9$, what should happen? Using `floor`, `ceil`, or `fix` may be useful here. The output of the function will be the calculated value of the Taylor polynomial $T_n(x)$.

Plot the Taylor polynomials for $x \in [-10, 10]$ for $n=4$, $n=j$, $n=k$, and $n=24$ using the above function `mycos` along with $y=\cos(x)$ on the same figure with the following specifications.

- The values of j and k will be random integers between 6 and 12 determined using the `randi` command. Use a `while` loop to make sure $k \neq j$.
- Have $y=\cos(x)$ dotted and in black, and have the others in different colors and/or shapes (dotted, dashed, etc.). Use your own judgment on `LineWidth`, etc. to make the graphs clear.
- Have an appropriate legend and title. (`sprintf` may be useful to put the values of n in the legend!)
- The vertical axis should only be between -2 and 2.

10. This problem is looking at Taylor/Maclaurin series and Taylor polynomials for $f(x)=\tan^{-1}(x)$.

$$\tan^{-1}(x) = \sum_{k=1}^{\infty} \frac{(-1)^{(k-1)}\, x^{2k-1}}{2k-1} \overset{\text{or}}{=} \sum_{k=0}^{\infty} \frac{(-1)^{k}\, x^{2k+1}}{2k+1}, \qquad x \in [-1, 1].$$

Create a function called `myatan.m` that calculates the nth-degree Taylor polynomial for $f(x)=\tan^{-1}(x)$ using the above Maclaurin series for $\tan^{-1}(x)$. The function should have two inputs; x and n. Make sure that n is a **nonnegative integer** using your function `isnnInt`; if not, an appropriate error message is displayed using the `error` command). The calculations should be such that x can be a number, vector, or matrix. The input n is the **highest degree term that will appear in the polynomial**. Thus if $n=8$, the degree of the polynomial will be no bigger than 8. Notice that if we enter $n=8$, we should get the same as $n=7$. Use the `warning` command if an odd value of n is entered. Careful: if we want $n=7$, what should k equal for the last term in the sum? If we input $n=8$, what should happen? Using `floor`, `ceil`, or `fix` may be useful here. The output of the function will be the calculated value of the Taylor polynomial $T_n(x)$.

Plot the Taylor polynomials for $x \in [-1.5, 1.5]$ for $n=3, 5, 13$, and 52 along with $y=\tan^{-1}(x)$ (`atan`) using the above function `myatan` on the same figure with the following specifications.

- Have $y = \tan^{-1}(x)$ dotted and in black, and have the others in different colors and/or shapes (dotted, dashed, etc.). Use your own judgment on `LineWidth`, etc. to make the graphs clear.
- Have an appropriate legend and title. (`sprintf` may be useful to put the values of n in the legend!)
- The vertical axis should only be between -2 and 2.

11. Taylor series and polynomials for $g(x) = \tan^{-1}(x^3)$.

(a) Using the Maclaurin series for $\tan^{-1}(x)$, find the Maclaurin series for $\tan^{-1}(x^3)$ and its radius of convergence, showing any work needed on paper. State the formula along with its radius of convergence.

(b) Integrate the series you get in part (a) for $a > 0$ to get a series for the following integral:

$$\int_0^a g(x)\,dx = \int_0^a \tan^{-1}(x^3)\,dx,$$

also find the radius of convergence for the series for this integral, showing all work on paper. State the formula for the series along with its the radius of convergence.

(c) Create a function called `atanint.m` that calculates the nth-degree Taylor polynomial for the series in part (b) for a given a and a given n. The function should check that $a > 0$, and that n is a nonnegative integer (use your function `isnnInt`); if not, appropriate error messages would be displayed using the `error` command. Just as in `myatan` above, the input n should be the highest degree that will appear in the partial sum (Taylor polynomial). Notice that the series is an alternating series. Using the alternating series estimation theorem, figure out what the estimated error would be for the partial sum with input n, and take that as the **second output**. The second output will be given whether or not it is asked for (such as the MATLAB command/function `size`).

(d) Use properties of integrals, the fact that $g(x)$ is either an even or odd function, and your function `atanint` if necessary to estimate the following (use `format long` for these answers). Show all work on paper.

i. $\int_0^{1/4} \tan^{-1}(x^3)\,dx$ (use $n = 5$ and $n = 20$),

ii. $\int_{-1/4}^{1/4} \tan^{-1}(x^3)\,dx$ (use $n = 5$ and $n = 20$),

iii. $\int_{-1/4}^{3/4} \tan^{-1}(x^3)\,dx$ (use $n = 5$ and $n = 20$).

Make sure your answers are clearly labeled.

CHAPTER 11

Numerical Integration

11.1. Approximating integrals/numerical integration

Our goal is to discuss *basic* ways to approximate the integral

$$\int_a^b f(x)\,dx.$$

The idea is that we view the integral as area "under the curve" (between the curve and the x-axis, with negative area if the curve is below the x-axis). We approximate the actual area (integral) by approximating the area with areas of regions that are easier to calculate. The key for any of numerical integration techniques is to subdivide the interval $[a, b]$ into n subintervals. In this document we are making our lives easier by dividing the interval into n subintervals of equal width:

$$\Delta x = \frac{b-a}{n}.$$

Fancier methods would use different width subintervals depending on how the function behaves on the domain, etc. Then n regions are formed based on the subdivisions and the values of $f(x)$, and the areas of these regions are calculated and totaled. The idea is that, as $n \to \infty$ or, equivalently, as the width $\Delta x \to 0$, these approximations approach the value of $\displaystyle\int_a^b f(x)\,dx$ (see Fig. 11.1).

The subdivision of the interval $[a, b]$ is such that $a = x_0$ and $b = x_n$ to get

$$a = x_0 < x_1 < x_2 < \cdots < x_{n-1} < x_n = b.$$

The formula for the points x_k is

$$x_k = a + k\Delta x, \quad k = 0, 1, \ldots n.$$

11.2. Riemann sums

Riemann sums are used to approximate $\int_a^b f(x)\,dx$ by using the areas of rectangles or trapezoids for the approximating areas. Each rectangle/trapezoid has width Δx. How we choose the height of the rectangles gives us different methods of approximation, and there is also the trapezoidal method.

Programming Mathematics Using MATLAB®
https://doi.org/10.1016/B978-0-12-817799-0.00017-X

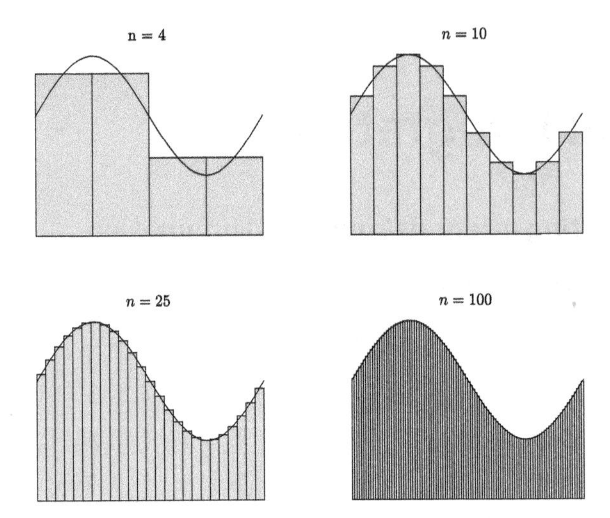

Figure 11.1 Approximating areas.

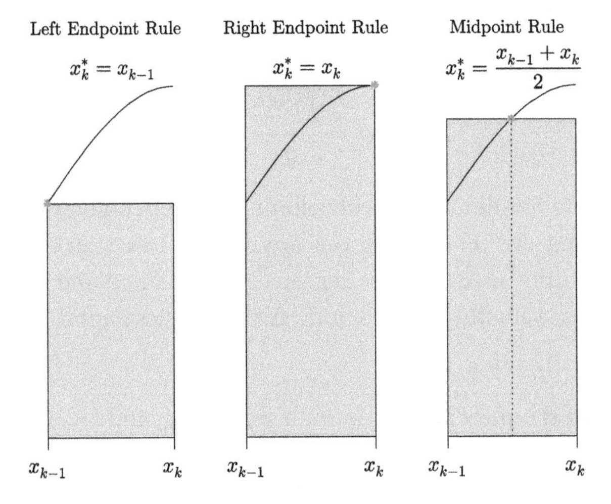

Figure 11.2 Sample points for approximating rectangles

For the kth subinterval $[x_{k-1}, x_k]$, we choose a *sample point* $x_k^* \in [x_{k-1}, x_k]$. The height of the kth rectangle is $f\left(x_k^*\right)$. Thus the general form of using rectangles for approximating $\int_a^b f(x)\,dx$ is

$$\int_a^b f(x)\,dx \approx \sum_{k=1}^{n} f\left(x_k^*\right) \Delta x.$$

The three most common methods using rectangles are by using the left endpoint, right endpoint, or midpoint of the subinterval to choose x_k^*, and thus the height of the rectangle (see Fig. 11.2 above).

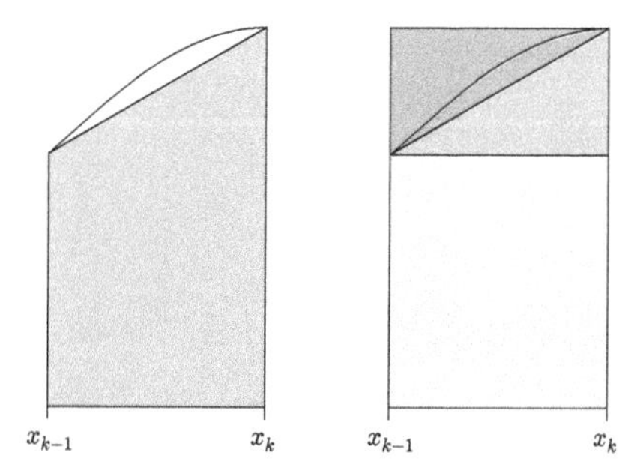

Figure 11.3 Trapezoidal rule.

If you look closely at Fig. 11.1, you will notice that the rectangles were drawn using the midpoint method.

The **trapezoidal rule** uses trapezoids instead of rectangles for the regions to approximate the area. The vertices of the kth trapezoid are $(x_{k-1}, 0)$, $(x_{k-1}, f(x_{k-1}))$, $(x_k, f(x_k))$, $(x_k, 0)$. The area of the kth trapezoid is

$$A_k = \frac{\Delta x}{2}\left(f(x_{k-1}) + f(x_k)\right).$$

Why? Because it is actually the average of the left and right endpoint rules (see Fig. 11.3 above).

Thus the formula for using the trapezoidal rule with n rectangles is

$$\begin{aligned}\int_a^b f(x)\,dx \approx T_n &= \frac{\Delta x}{2}\sum_{k=1}^{n}\left(f(x_{k-1}) + f(x_k)\right)\\ &= \frac{\Delta x}{2}\left(f(x_0) + 2f(x_1) + 2f(x_2) + \cdots + 2f(x_{n-1}) + f(x_n)\right)\end{aligned}$$

Fig. 11.4 shows all four Riemann sums.

11.3. Error bounds

We have error bounds for the midpoint and trapezoidal rules:

Theorem 11.3.1 (Error bounds). Suppose $|f''(x)| \le m$ for all $x \in [a, b]$. Then using the midpoint rule to estimate the integral $\int_a^b f(x)\,dx$ will have an error ε_M with

$$|\varepsilon_M| \le \frac{m(b-a)^3}{24n^2}$$

and using the trapezoidal rule will have an error ε_T with

$$|\varepsilon_T| \le \frac{m(b-a)^3}{12n^2}.$$

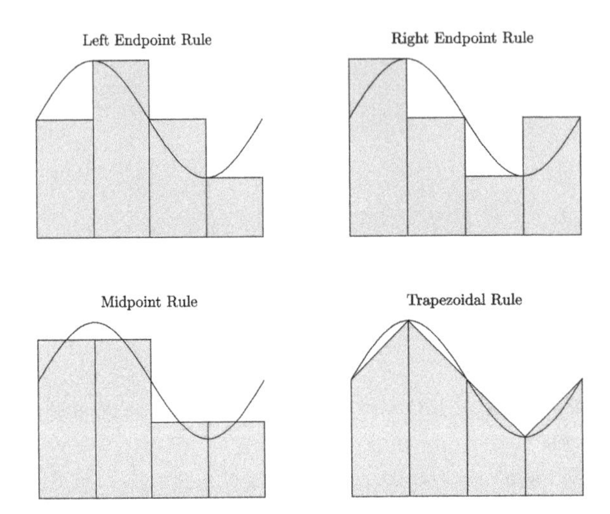

Figure 11.4 Common Riemann sums.

We will not prove these error bounds in these notes; that is for a different course. Notice that the error bounds get smaller faster for the midpoint rule than the trapezoidal rule (which may seem counterintuitive).

11.4. Simpson's rule

For the Simpson rule of approximating integrals, areas under quadratics are used to approximate the area under the curve, $A = \int_a^b f(x)\,dx$. The Simpson rule starts by subdividing the interval $[a, b]$ into n equal subintervals of width $\Delta x = \frac{b-a}{n}$, where n **is even**. Then we take *pairs of consecutive subintervals* and estimate the area under the curve with a quadratic (see Fig. 11.5). Thus our first approximating quadratic is estimating the area from x_0 to x_2, then the second approximating quadratic is estimating the area from x_2 to x_4, etc., until the $n/2$th approximating quadratic is estimating the area from x_{n-2} to x_n. Notice that our formulas for x_0, x_1, x_k, etc. are as in the previous rules.

It is done this way because in order to come up with a quadratic

$$A_1x^2 + B_1x + C_1,$$

one needs three points on the quadratic. Having these pairs of subintervals allows for three data points at each approximation. Let us investigate the first estimating quadratic; we want to estimate the area under the curve $y = f(x)$ from x_0 to x_2. We have three data points: $(x_0, f(x_0))$, $(x_1, f(x_1))$, and $(x_2, f(x_2))$. For ease of notation, let us have $y_0 = f(x_0)$, etc.

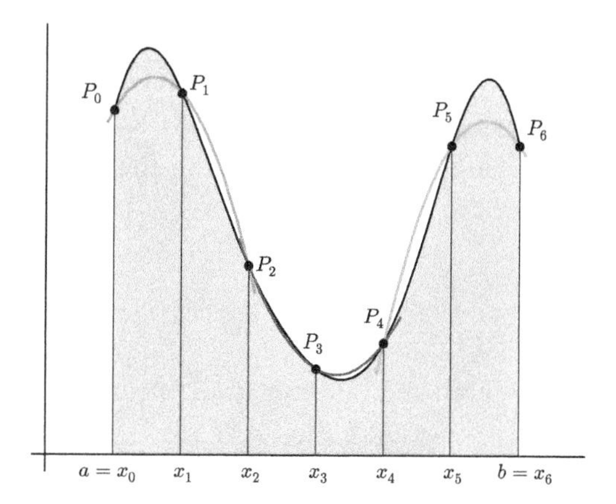

Figure 11.5 Simpson's rule.

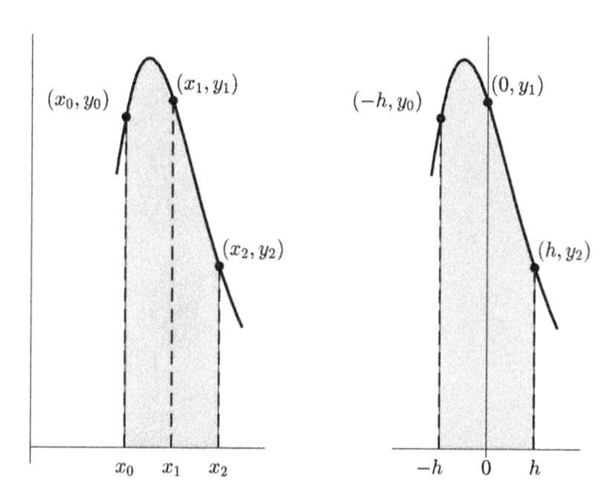

Figure 11.6 Translating areas.

In order to make finding the quadratic $A_1x^2 + B_1x + C_1$ easier, let us shift the x's to be from $-h$ to h. So $x_0 \to -h$, $x_1 \to 0$, and $x_2 \to h$, with $h = \Delta x$ (see Fig. 11.6).

Notice that the area under the curve does not change and we now have the data points $(-h, y_0)$, $(0, y_1)$, and (h, y_2) to figure out the approximating quadratic and/or approximating area.

To figure out $A_1x^2 + B_1x + C_1$:

$$x = 0 \implies C_1 = y_1, \tag{11.1}$$

$$x = -h \implies A_1h^2 - B_1h + C_1 = y_0, \tag{11.2}$$

$$x = h \implies A_1h^2 + B_1h + C_1 = y_2. \tag{11.3}$$

Adding Eqs. (11.2) and (11.3), we see that

$$y_0 + y_2 = 2A_1h^2 + 2C_1. \tag{11.4}$$

If we were actually going to figure out the coefficients of the approximating quadratic, we have already figured C_1 in (11.1), and we can use that in (11.4) to figure out A_1, and then use either (11.2) or (11.3) to find B_1.

But we actually do not need to figure out all of the coefficients because all we need is to figure out the area under the quadratic, i.e.,

$$\int_{-h}^{h} A_1x^2 + B_1x + C_1\, dx.$$

Using facts about integrals of even and odd functions, we get

$$\begin{aligned}\int_{-h}^{h} A_1x^2 + B_1x + C_1\, dx &= 2\int_{0}^{h} A_1x^2 + C_1\, dx \\ &= \frac{2}{3}A_1x^3 + 2C_1x\Big|_0^h \\ &= \frac{2}{3}A_1h^3 + 2C_1h \\ &= \frac{h}{3}\left(2A_1h^2 + 6C_1\right). \qquad (11.5)\end{aligned}$$

By combining (11.1) with (11.4), we get

$$y_0 + 4y_1 + y_2 = 2A_1h^2 + 6C_1. \qquad (11.6)$$

Thus

$$\int_{-h}^{h} A_1x^2 + B_1x + C_1\, dx = \frac{h}{3}\left(y_0 + 4y_1 + y_2\right).$$

Now if we shift the x back, the area under the curve remains the same and we get

$$\int_{x_0}^{x_2} A_1x^2 + B_1x + C_1\, dx = \frac{h}{3}\left(y_0 + 4y_1 + y_2\right).$$

Likewise

$$\int_{x_2}^{x_4} A_2x^2 + B_2x + C_2\, dx = \frac{h}{3}\left(y_2 + 4y_3 + y_4\right).$$

Thus the general formula for the Simpson rule is

$$\boxed{\int_a^b f(x)\, dx \approx \frac{h}{3}\left(y_0 + 4y_1 + 2y_2 + 4y_3 + 2y_4 + \cdots + 2y_{n-2} + 4y_{n-1} + y_n\right).}$$

Notice the pattern of the coefficients: $1, 4, 2, 4, 2, \ldots, 2, 4, 1$.

There is also an *error estimate* for the Simpson rule (not proven here).

Theorem 11.4.1 (Simpson's rule error estimate). If $|f^{(4)}(x)| \leq m$ for $x \in [a, b]$, then using Simpson's rule to estimate $\int_a^b f(x)\,dx$ with n subintervals will have an error ε_S where

$$|\varepsilon_S| \leq \frac{m(b-a)^5}{180n^4}.$$

11.5. Exercises

Note that all ERROR CHECKS within your functions should use the `error` command and display a meaningful error message.

1. Create a function `RSum` that takes as input a function `fstring`, `a`, `b`, `n`, `sumType`, and `color`. The commands `syms`, `str2sym`, `subs`, and `double` may be useful or necessary. The `switch` selection statement may also come in handy for this assignment.
   - The input `fstring` will be a string for the function $f(x)$ that we want to integrate numerically. The string is in single quotes using standard MATLAB® notation WITHOUT component-wise calculations. No error checking will be done on this.
   - The inputs `a` and `b` are the lower and upper bounds, respectively for the integration. An error check will check that `a` < `b`; if not, an appropriate error message will be displayed.
   - The input `n` will be the number of rectangles/subintervals that will be made for the numerical integration (Riemann Sum). An error check will check that `n` is a positive integer; if not, an appropriate error message will be displayed.
   - The input `sumType` will determine which type of Riemann sum will be calculated and plotted. It will be a string in single quotes and will be either `'right'`, `'left'`, `'mid'` or `'trap'`; if not, an appropriate error message will be displayed.
   - The input `color` will specify which color will be used to fill in the rectangles. It could be a vector or a string name for a color. No error checking will be done on this input.

   The function will have 1 output; the calculated Riemann sum (using the right Riemann sum if `sumType` = `'right'`, the left Riemann sum if `sumType` = `'left'`, the midpoint rule if `sumType` = `'mid'`, and the trapezoidal rule if `sumType` = `'trap'`). The value that is the output should NOT be symbolic—thus you may need to use the `double` command.

   The function will also plot the rule (depending on the value of `sumType`) of $y = f(x)$ from $x = a$ to $x = b$ using n subdivisions. You will use the `fill` command. Both the edges of the rectangles/trapezoids and the function $y = f(x)$ will be plotted in black.

There will be no titles, axis labels, or other axis commands within the function; the user can add these outside of the function.

2. For $f(x) = \sin^4(\pi x) + 2x$, figure out the following by hand on paper and/or using MATLAB to help with some calculations. Any calculations not done by hand should be shown in MATLAB ("show your work").
   - **(a)** Approximate $\int_1^3 f(x)\,dx$ using a left Riemann sum and $n = 4$.
   - **(b)** Approximate $\int_1^3 f(x)\,dx$ using a right Riemann sum and $n = 4$.
   - **(c)** Approximate $\int_1^3 f(x)\,dx$ using a midpoint rule and $n = 4$.
   - **(d)** Approximate $\int_1^3 f(x)\,dx$ using a trapezoidal rule and $n = 4$.
3. **(a)** Use the `subplot` command and your function to show all four Riemann sums with $n = 4$ on the same figure using your `RSum` function for $\int_1^3 f(x)\,dx$ using $f(x) = \sin^4(\pi x) + 2x$. This will be a 2×2 figure where the top row will be the left and right Riemann sums and the second row will be the midpoint rule and trapezoidal rule. Let the left Riemann sum be in MyGreen ([0, 0.4078, 0.3412]), the right Riemann sum in yellow, the midpoint rule in red with `alpha(0.3)`, and the trapezoidal rule in gray ([0.75, 0.75, 0.75]). Make sure you add titles specifying which rule is which.
   (b) Compare the answers you get from your function with your answers in the previous problem. Are they what you expected?
4. Use the `subplot` command to plot the one of the rules (your choice for the color and rule but use the same for all four within these subplots) using your `RSum` function on a figure with 2×2 subplots: $n = 4$, $n = 8$, $n = 30$, $n = 75$ for $\int_1^3 (\sin^4(\pi x) + 2x)\,dx$.
5. Create a function `SRule` that takes as input a function `fstring`, `a`, `b`, and `n`.
   - The input `fstring` is a string for the function $f(x)$ in single quotes using standard MATLAB notation WITHOUT component-wise operations, just as in `RSum`. No error check is done on this input.
   - The inputs `a` and `b` are the lower and upper bounds, respectively for the integration. An error check will check that `a <b`; if not, an appropriate error message will be displayed.
   - The input `n` will be the number of rectangles/subintervals that will be made for the Simpson rule. An error check will check that `n` is a positive EVEN integer; if not, an appropriate error message will be displayed.

   The commands `syms`, `str2sym`, `sum`, `subs`, and `double` will be useful or necessary. The output of the function will be the approximation of the integral of $f(x)$ from a

to b using the Simpson rule on the given n. Just as in the `RSum` function, it should return a numerical approximation so you may need to use the `double` command.

6. Create a function `SPlot.m` that has the same inputs (and error checks) as `SRule` in the above exercise. This function will plot the Simpson rule of $y = f(x)$, from $x = a$ to $x = b$ using n subdivisions to create a figure similar to Fig. 11.5. You will use the `syms`, `str2sym`, `polyfit`, and the `polyval` commands. The approximating quadratics will be plotted without a color specified (so that the graphs cycle through the colors), and the curve $y = f(x)$ will be in black. Have the domains in the plots for the approximating quadratics be 0.05 beyond the x_k used in the approximations. For example, if the first quadratic is approximating the curve from 1 to 1.5, then have the plot of the quadratic be from 0.95 to 1.55. You will also plot the subdivisions in black; these will be vertical lines from $(x_k, 0)$ to (x_k, y_k) for each of $x_0, x_1, \ldots, x_n$.
7. We will investigate Simpson's rule to estimate $\int_1^3 f(x)\,dx$ for $f(x) = \sin^4(\pi x) + 2x$, above with $n = 2$ and $n = 4$. For the subdivisions, figure out what the FIRST approximating quadratics would be. This should be done on paper and turned in, showing all work, explaining it clearly, and using exact values. Any calculations using technology should be done in MATLAB. The answers for the approximating quadratics (and the subintervals they are for) should be written clearly. Use the interval $[-h, h]$ and figure out the coefficients A, B, and C based on the values of y_k, y_{k+1}, and y_{k+2} from the notes. That quadratic is based on the middle x-value equaling 0. Use a horizontal shift so that the middle value is now at x_{k+1}.
   (a) Figure out the FIRST approximating quadratic for $n = 2$.
   (b) Figure out the FIRST approximating quadratic for $n = 4$.
8. Check your answers in the above problem using the command `polyfit` and even `poly2sym`, clearly labeling your answers.
   (a) Check for $n = 2$.
   (b) Check for $n = 4$.
9. Use your `SRule` function to approximate $\int_1^3 (\sin^4(\pi x) + 2x)\,dx$ with the Simpson rule using $n = 2$, $n = 4$, $n = 8$, and $n = 16$. Also, use `subplot` as in previous exercises with your `SPlot` to visualize the Simpson rule for these four numerical approximations.

CHAPTER 12

The Gram–Schmidt Process

12.1. General vector spaces and subspaces

12.1.1 Vector spaces

Definition 12.1.1. A **vector space** is a set V of elements called **vectors** that have operations called vector addition and scalar multiplication defined so that the following conditions hold for any $\mathbf{u}, \mathbf{u}, \mathbf{w} \in V$ and scalars a, and b.

- **Closure properties**
  c-1. $\mathbf{u} + \mathbf{v} \in V$
  c-2. $a\mathbf{u} \in V$
- **Addition properties**
  a-1. $\mathbf{u} + \mathbf{v} = \mathbf{v} + \mathbf{u}$ (commutativity)
  a-2. $\mathbf{u} + (\mathbf{v} + \mathbf{w}) = (\mathbf{u} + \mathbf{v}) + \mathbf{w}$ (associativity)
  a-3. $\exists$ zero vector $\mathbf{0} \in V$ such that (additive identity)
  $$\mathbf{u} + \mathbf{0} = \mathbf{u} \quad \forall \mathbf{u} \in V$$
  a-4. $\forall \mathbf{u} \in V, \exists -\mathbf{u} \in V$ such that (additive inverse)
  $$\mathbf{u} + (-\mathbf{u}) = \mathbf{0}$$
- **(Scalar) multiplication properties**
  m-1. $a(b\mathbf{u}) = (ab)\mathbf{u}$ (associativity)
  m-2. $a(\mathbf{u} + \mathbf{v}) = a\mathbf{u} + a\mathbf{v}$ (distributive)
  m-3. $(a + b)\mathbf{u} = a\mathbf{u} + b\mathbf{u}$ (distributive)
  m-4. $1\mathbf{u} = \mathbf{u} \quad \forall \mathbf{u} \in V$

Scalars are commonly the set of real numbers (called a **real vector space**) or the set of complex numbers (called a **complex vector space**).

Examples of vector spaces

1. $\mathbb{R}^n$: Euclidean vector space of $\mathbb{R}^n$
2. M_{mn}: the set of $m \times n$ matrices.
3. The set of all functions with domain $\mathbb{R}$:
   - Addition is defined *pointwise*: $f + g = h$ is the function $h(x) = (f + g)(x) = f(x) + g(x)$.
   - af is defined as the function $af(x)$.

Programming Mathematics Using MATLAB®
https://doi.org/10.1016/B978-0-12-817799-0.00018-1

Note: you could define a different domain instead of $\mathbb{R}$. You could also change it to be all *continuous* functions on domain D, all *differentiable* functions with domain D, etc.

4. P_n: the set of all real-polynomials of degree $\leq n$.
5. $C^n[a, b]$: the set of all functions whose nth derivatives exist and are continuous on $[a, b]$ (thus the function, first derivative, second derivative, ..., nth derivative are all continuous on $[a, b]$).
6. The set of all functions that satisfy the differential equation $3y'' - y' + y = 0$.

Theorem 12.1.1 (Properties of vectors). Let V be a vector space, $\mathbf{v} \in V$, $\mathbf{0} \in V$ and 0 the zero scalar. Then

1. $0\mathbf{v} = \mathbf{0}$, and $c\mathbf{0} = \mathbf{0}$ for any scalar c.
2. $(-1)\mathbf{v} = -\mathbf{v}$.
3. If $c\mathbf{v} = \mathbf{0}$, then either $c = 0$ or $\mathbf{v} = \mathbf{0}$.

12.1.2 Subspaces

Definition 12.1.2. Let V be a vector space and U be a nonempty subset of V. If U is a vector space under the same addition and scalar multiplication, then U is called a **subspace** of V.

U is a subspace if it is closed under addition and scalar multiplication. All other vector space properties in the definition are inherited from V.

Example 12.1.1. Which of the following are subspaces of M_{nn}?

(a) The subset of all symmetric matrices.
(b) The subset of all matrices that are not symmetric.
(c) The subset of invertible matrices.
(d) All diagonal matrices.

Example 12.1.2. Which subsets W are subspaces of the vector space V?

(a) $V = P_3$, $W = P_2$.
(b) $V = P_3$, $W = \{p(x) \mid p(x) = ax^2 + bx + 1,\ a, b \in \mathbb{R}\}$.
(c) $V =$ set of all functions with domain being $\mathbb{R}$,
$W = \{f(x) \in V \mid f(0) = 0\}$.
(d) $V =$ set of all functions with domain being $\mathbb{R}$,
$W = \{f(x) \in V \mid f(0) = 1\}$.
(e) $V =$ set of all functions with domain being $\mathbb{R}$,
$W = \{f(x) \in V \mid \int_a^b f(x)\, dx = 0\}$ for some fixed $a, b \in \mathbb{R}$.

12.2. Linear combinations of vectors

Definition 12.2.1. Let V be a vector space and $\mathbf{v}_1, \mathbf{v}_2, \ldots, \mathbf{v}_m \in V$. We say $\mathbf{v} \in V$ is a **linear combination** of $\mathbf{v}_1, \mathbf{v}_2, \ldots, \mathbf{v}_m$ if there exist scalars $a_1, a_2, \ldots a_m$ such that

$$\mathbf{v} = a_1\mathbf{v}_1 + a_2\mathbf{v}_2 + \cdots + a_m\mathbf{v}_m.$$

Example 12.2.1. Is the first vector a linear combination of the other vectors? If so, write the linear combination.

(a) $\mathbf{x} = (-11, 9, -3)$; $\mathbf{v}_1 = (1, -1, 1), \mathbf{v}_2 = (2, 1, 4), \mathbf{v}_3 = (-2, 3, 1)$.

(b) $\mathbf{x} = \begin{bmatrix} 1 & -25 \\ 8 & -13 \end{bmatrix}$; $\mathbf{v}_1 = \begin{bmatrix} 2 & 0 \\ 1 & -1 \end{bmatrix}, \mathbf{v}_2 = \begin{bmatrix} 0 & 1 \\ 3 & 4 \end{bmatrix}, \mathbf{v}_3 = \begin{bmatrix} 1 & 5 \\ -1 & 2 \end{bmatrix}$.

(c) $f(x) = 2x^2 + x - 3$; $g(x) = x^2 - x + 1, h(x) = x^2 + 2x - 2$.

Definition 12.2.2. Let V be a vector space and $\mathbf{v}_1, \mathbf{v}_2, \ldots, \mathbf{v}_m \in V$. These vectors **span** V if every vector V can be written as a linear combination of $\mathbf{v}_1, \mathbf{v}_2, \ldots, \mathbf{v}_m$.

Example 12.2.2. The following polynomials span $V = P_2$:

$$f_1 = 1, f_2 = x, f_3 = x^2 + x, f_4 = x^2 - 1.$$

Example 12.2.3. The following matrices span $V = M_{22}$:

$$A_1 = \begin{bmatrix} 1 & 0 \\ 0 & 0 \end{bmatrix}, A_2 = \begin{bmatrix} 1 & 1 \\ 0 & 0 \end{bmatrix}, A_3 = \begin{bmatrix} 0 & 0 \\ 1 & 0 \end{bmatrix}, A_4 = \begin{bmatrix} 0 & 0 \\ 0 & 1 \end{bmatrix}.$$

Theorem 12.2.1. Let $S = \{\mathbf{v}_1, \mathbf{v}_2, \ldots \mathbf{v}_m\} \subset V$. Let U be the set consisting of all linear combinations of vectors from S. Then U is a subspace of V spanned by the vectors in S, and we say U is the vector space *generated by* S, or is the *span of* S. We denote this by

$$U = \text{span}\, S = \text{span}\{\mathbf{v}_1, \mathbf{v}_2, \ldots \mathbf{v}_m\}.$$

Example 12.2.4. Determine whether the vector $\mathbf{v}$ is in the span of S.

(a) $\mathbf{v} = (1, 4, -3)$, $S = \{(1, 0, 1), (1, 1, 0), (3, 1, 2)\}$ (yes).

(b) $\mathbf{v} = (1, 1, 2)$, $S = \{(0, 1, 0), (3, 5, 6), (1, 2, 1)\}$ (no).

Example 12.2.5. Give examples of three other functions in the subspace span S, where

$$S = \{2x + 1,\ 3x^2 + x - 3\}.$$

12.3. Linear independence and bases

12.3.1 Linear independence

Definition 12.3.1. Let V be a vector space.

1. $S = \{\mathbf{v}_1, \mathbf{v}_2, \ldots, \mathbf{v}_m\} \subset V$ is **linearly dependent** if there exists a nontrivial solution to

$$c_1\mathbf{v}_1 + c_2\mathbf{v}_2 + \cdots + c_m\mathbf{v}_m = \mathbf{0}. \qquad (\star)$$

2. S is LINEARLY INDEPENDENT if the only solution to ($\star$) is the trivial solution

$$c_1 = c_2 = \cdots = c_m = 0.$$

Example 12.3.1. Are the following sets of vectors linearly independent or linearly dependent?

(a) $\{1, x, x^2\} \subset P_2$.

(b) $\{x^2 + x^3\} \subset P_3$.

(c) $\left\{\begin{bmatrix} 1 & 0 \\ 0 & 1 \end{bmatrix}, \begin{bmatrix} 0 & 0 \\ 0 & 1 \end{bmatrix}\right\} \subset M_{22}$.

(d) $\{f_1, f_2, f_3, f_4\} \subset P_3$ where

$$f_1 = 1, f_2 = x, f_3 = x^3 + x, \text{ and } f_4 = x^2 - 1.$$

(e) $\{f_1, f_2, f_3\} \subset P_1$ where

$$f_1 = 1, f_2 = x, \text{ and } f_3 = 2x - 1.$$

Theorem 12.3.1. Let V be a vector space and $V = \text{span}\{\mathbf{v}_1, \mathbf{v}_2, \ldots \mathbf{v}_m\}$. Each vector in V can be expressed uniquely as a linear combination of these vectors iff the vectors are linearly independent.

Proof. ($\Longleftarrow$) Assume the set of vectors $S = \{\mathbf{v}_1, \mathbf{v}_2, \dots \mathbf{v}_m\}$ are LI. Let $\mathbf{v} \in V$. Since $V =$ span S, by definition of span we can write $\mathbf{v}$ as a linear combo of these vectors. Suppose we have two ways of writing $\mathbf{v}$ as a linear combination:

$$\mathbf{v} = \sum a_k \mathbf{v}_k, \qquad \mathbf{v} = \sum b_k \mathbf{v}_k.$$

Then we have

$$\sum (a_k - b_k)\mathbf{v}_k = \mathbf{0}.$$

But since these vectors are LI, $a_k - b_k = 0$ for all k. Thus $a_k = b_k$ for all k, and there's only one way of expressing $\mathbf{v}$ as a linear combo of the vectors in S.

($\Longrightarrow$) Let $\mathbf{v} \in V$ be arbitrary and we are given that $\mathbf{v}$ can be written uniquely as a linear combo of vectors in S. Since $\mathbf{0} \in V$, then $\mathbf{0}$ can be written uniquely as a linear combination of vectors in S. We know

$$\mathbf{0} = \sum 0\mathbf{v}_k,$$

so this must be the only solution, thus the vectors in S are linear independent. □

12.3.2 Bases

Definition 12.3.2. Let V be a vector space. Let $S = \{\mathbf{v}_1, \mathbf{v}_2, \dots, \mathbf{v}_m\} \subset V$ be such that

1. S is a linearly independent set of vectors, and
2. $V = \text{span}\, S$.

Then S is called a BASIS for V.

Thus if S is a basis for V, each vector in V can be expressed uniquely as a linear combination of vectors in S.

Example 12.3.2. The standard basis for $\mathbb{R}^n$ is

$$\mathbf{e}_1 = (1, 0, \dots, 0),\ \mathbf{e}_2 = (0, 1, 0, \dots, 0),\ \dots,\ \mathbf{e}_n = (0, \dots, 0, 1).$$

Example 12.3.3. A basis for $V = M_{22}$ is

$$A_1 = \begin{bmatrix} 1 & 0 \\ 0 & 0 \end{bmatrix}, \quad A_2 = \begin{bmatrix} 0 & 1 \\ 0 & 0 \end{bmatrix}, \quad A_3 = \begin{bmatrix} 0 & 0 \\ 1 & 0 \end{bmatrix}, \quad A_4 = \begin{bmatrix} 0 & 0 \\ 0 & 1 \end{bmatrix}.$$

This is the standard basis for M_{22}.

Example 12.3.4. The standard basis for $V = P_3$ is

$$f_1 = 1, \quad f_2 = x, \quad f_3 = x^2, \quad f_4 = x^3.$$

> **Theorem 12.3.2.** Let $\{\mathbf{v}_1, \mathbf{v}_2, \ldots, \mathbf{v}_n\}$ be a basis for vector space V. If $\{\mathbf{w}_1, \mathbf{w}_2, \ldots, \mathbf{w}_m\}$ is a set of more than n vectors in V, then this set is linearly dependent.

Proof. Let $\sum c_k \mathbf{w}_k = \mathbf{0}$. Since $\{\mathbf{v}_1, \mathbf{v}_2, \ldots, \mathbf{v}_n\}$ is a basis for V, each of the vectors $\mathbf{w}_k$ can be written as a linear combo of these basis vectors.

$$\mathbf{w}_1 = a_{11}\mathbf{v}_1 + a_{12}\mathbf{v}_2 + \ldots$$

Thus we get

$$\sum c_k(a_{k1}\mathbf{v}_1 + a_{k2}\mathbf{v}_2 + \cdots + a_{kn}\mathbf{v}_n) = \mathbf{0}.$$

If we rearrange, we get

$$(c_1a_{11} + c_2a_{21} + \cdots c_ma_{m1})\mathbf{v}_1 + \cdots = 0.$$

Thus $a_{11}c_1 + a_{21}c_2 + \cdots + a_{m1}c_m = 0$, etc. Notice that we have m variables and n equations, but $n < m$ so at least one of the variables is free. Therefore there are nontrivial solutions for the c_k so the set is LD. □

> **Corollary 12.3.1.** All bases for a vector space V have the same number of vectors.

> **Definition 12.3.3.** The number of vectors in the basis of a vector space V is called the DIMENSION OF V and is denoted by $\dim(V)$.

Example 12.3.5. $\dim(M_{22}) = 4$, $\dim(P_2) = 3$.

Some vector spaces have infinite dimension!

Example 12.3.6. $P =$ set of all polynomials and $C^2[0, 1]$, etc. are infinite dimensional vector spaces.

Example 12.3.7. What are the dimensions of the following spaces?

(a) M_{44}.

(b) $V =$ set of all diagonal 4×4 matrices.

(c) $W =$ set of all upper triangular 4×4 matrices.

Theorem 12.3.3. **1.** If $V = \{\mathbf{0}\}$, then $\dim(V) = 0$

2. If $\dim(V) = 1$, $V \subset \mathbb{R}^n$, $(n = 2, 3)$, then V is a line through the origin.

3. If $\dim(V) = 2$, $V \subset \mathbb{R}^n$, $(n = 2, 3)$, then V is a plane through the origin.

Theorem 12.3.4. Let V be a vector space and $\dim(V) = n$.

Let $S = \{\mathbf{v}_1, \mathbf{v}_2, \ldots, \mathbf{v}_n\} \subset V$.

1. If S is a linearly independent set, then S is a basis for V.

2. If $\operatorname{span}(S) = V$, then S is a basis for V.

Example 12.3.8. $\begin{bmatrix} 1/\sqrt{2} \\ 0 \\ 1/\sqrt{2} \end{bmatrix}, \begin{bmatrix} -1/\sqrt{2} \\ 0 \\ 1\sqrt{2} \end{bmatrix}, \begin{bmatrix} 0 \\ 1 \\ 0 \end{bmatrix}$ are linearly independent in $\mathbb{R}^3$ (check it!) so the set of these vectors form a basis for $\mathbb{R}^3$.

Theorem 12.3.5. Let V be a vector space of dimension n. Let $S = \{\mathbf{v}_1, \mathbf{v}_2, \ldots, \mathbf{v}_m\} \subset V$. Suppose S is a linearly independent set with $m < n$. Then there exist vectors $\mathbf{v}_{m+1}, \mathbf{v}_{m+2}, \ldots, \mathbf{v}_n$ such that

$$S \cup \{\mathbf{v}_{m+1}, \mathbf{v}_{m+2}, \ldots, \mathbf{v}_n\}$$

form a basis for V.

Example 12.3.9. $\begin{bmatrix} 1 & 0 \\ 0 & 0 \end{bmatrix}, \begin{bmatrix} 0 & 0 \\ 0 & 1 \end{bmatrix}$ form a linearly independent set in the space of lower triangular matrices. We need one more to form a basis:

$$\begin{bmatrix} 0 & 0 \\ 5 & 0 \end{bmatrix}.$$

12.4. Rank

Definition 12.4.1. Let A be an $m \times n$ matrix. The rows of A may be viewed as row vectors $R = \{r_1, \ldots, r_m\}$, and the column vectors $C = \{c_1, \ldots c_n\}$. Each row vector has n components, so $R \subset \mathbb{R}^n$, and each column vector has m components, so $C \subset \mathbb{R}^m$. The span(R) is a subspace of $\mathbb{R}^n$ and is called the ROW SPACE OF A, and span(C) is a subspace of $\mathbb{R}^m$ and is called the COLUMN SPACE OF A.

Theorem 12.4.1. The row space and column space of a matrix A have the same dimension (even if these subspaces live in different vector spaces).

Definition 12.4.2. The dimension of the row space and the column space of a matrix A is called the RANK OF A, denoted by rank(A).

$$\text{rank}(A) = \dim(\text{row space of } A) = \dim(\text{col space of } A).$$

How do we find the rank and dimension of these spaces?

Theorem 12.4.2. Let A and B be row equivalent matrices. Then A and B have the same row space, and $\text{rank}(A) = \text{rank}(B)$.

Proof. Since A and B are row equivalent, the rows of B can be obtained from the rows of A through elementary row operations, and vice versa. Thus each row of B is a linear combination of the rows of A. Thus the row vectors of B are in the row space of A, and vice versa. Thus the row spaces are the same. □

Theorem 12.4.3. (Really a corollary) Let A be a matrix in reduced echelon form. The nonzero row vectors of A are a basis for the rowspace of A. Thus the number of nonzero rows in A are rank(A).

Theorem 12.4.4. (Again, really a corollary) Let $E = \text{rref}(A)$. The nonzero row vectors of E form a basis for the row space of A. The number of nonzero rows in E is rank(A).

Corollary 12.4.1. To find a basis for the column space of A, put A^T in reduced echelon form.

Theorem 12.4.5. Let $A\mathbf{x} = \mathbf{y}$ be an $m \times n$ system and let B be the augmented matrix $[A\,\mathbf{y}]$.

1. If $\text{rank}(A) = \text{rank}(B) = r$ and $r = n$, then the solution is unique.
2. If $\text{rank}(A) = \text{rank}(B) = r$ and $r < n$, then the solution has infinitely many solutions.
3. If $\text{rank}(A) \neq \text{rank}(B)$, then there are no solutions.

Theorem 12.4.6. (Summary) Let A be an $n \times n$ matrix. TFAE

1. $\det(A) \neq 0$.
2. A is nonsingular.
3. A is invertible.
4. A is row equivalent to I_n.
5. The system of equations $A\mathbf{x} = \mathbf{b}$ has a unique solution (no matter what the value of $\mathbf{b}$ is).
6. $\text{rank}(A) = n$.
7. The column vectors of A are linearly independent.
8. The column vectors of A span $\mathbb{R}^n$.
9. The column vectors of A form a basis for $\mathbb{R}^n$.

12.5. Orthonormal vectors and the Gram–Schmidt process

12.5.1 Orthogonal and orthonormal vectors

Recall the following definitions.

Definition 12.5.1. The DOT PRODUCT or INNER PRODUCT or SCALAR PRODUCT of $\mathbf{u} = (u_1, u_2, \ldots, u_n)$ and $\mathbf{v} = (v_1, v_2, \ldots, v_n)$ in $\mathbb{R}^n$, is

$$\mathbf{u} \cdot \mathbf{v} = u_1 v_1 + u_2 v_2 + \cdots + u_n v_n = \sum_{k=1}^{n} u_k v_k.$$

Definition 12.5.2. The NORM/MAGNITUDE/LENGTH of a vector $\mathbf{u} = (u_1, u_2, \ldots, u_n) \in \mathbb{R}^n$ is

$$\|\mathbf{u}\| = \sqrt{u_1^2 + u_2^2 + \cdots + u_n^2} = (\mathbf{u} \cdot \mathbf{u})^{1/2}.$$

Any vector $\mathbf{u} \neq \mathbf{0}$ can be ***normalized***; i.e., a unit vector $\mathbf{v}$ can be found in the same direction as $\mathbf{u}$:

$$\mathbf{v} = \frac{1}{\|\mathbf{u}\|}\mathbf{u}.$$

Definition 12.5.3. A pair of vectors $\mathbf{u}$, $\mathbf{v}$ in $\mathbb{R}^n$ is ORTHOGONAL if $\mathbf{u} \cdot \mathbf{v} = 0$.

(In $\mathbb{R}^2$ and $\mathbb{R}^3$, orthogonal is synonymous with perpendicular.)

Definition 12.5.4. A set of vectors $S \subset V$ is said to be an ORTHOGONAL SET if every pair of vectors in the set is orthogonal. The set S is an ORTHONORMAL SET if is an orthogonal set and every vector is a unit vector.

Example 12.5.1. $S = \left\{ \mathbf{u}_1 = \begin{bmatrix} 1 \\ -1 \\ 1 \end{bmatrix}, \mathbf{u}_2 = \begin{bmatrix} 1 \\ 2 \\ 1 \end{bmatrix}, \mathbf{u}_3 = \begin{bmatrix} 1 \\ 0 \\ -1 \end{bmatrix} \right\}$,

$$\mathbf{u}_1 \cdot \mathbf{u}_2 = 0,$$
$$\mathbf{u}_1 \cdot \mathbf{u}_3 = 0,$$
$$\mathbf{u}_2 \cdot \mathbf{u}_3 = 0,$$

$\Longrightarrow$ S is an orthogonal set.

Example 12.5.2. The vectors $\mathbf{v}_1 = \begin{bmatrix} \frac{1}{\sqrt{2}} \\ -\frac{1}{\sqrt{2}} \\ 0 \end{bmatrix}$, $\mathbf{v}_2 = \begin{bmatrix} \frac{1}{\sqrt{2}} \\ \frac{1}{\sqrt{2}} \\ 0 \end{bmatrix}$, $\mathbf{v}_3 = \begin{bmatrix} 0 \\ 0 \\ -1 \end{bmatrix}$ form an orthonormal set.

Theorem 12.5.1. An orthogonal set is linearly independent.

Proof. Let $S = \{\mathbf{v}_1, \mathbf{v}_2, \ldots, \mathbf{v}_m\}$ be an orthogonal set of nonzero vectors in V. Consider the equation

$$c_1\mathbf{v}_1 + c_2\mathbf{v}_2 + \cdots + c_m\mathbf{v}_m = \mathbf{0}.$$

Take the equation and take the dot product with a vector $\mathbf{v}_k \in S$:

$$(c_1\mathbf{v}_1 + c_2\mathbf{v}_2 + \cdots + c_m\mathbf{v}_m) \cdot \mathbf{v}_k = 0 \cdot \mathbf{v}_k,$$
$$c_1\mathbf{v}_1 \cdot \mathbf{v}_k + c_2\mathbf{v}_2 \cdot v_k + \cdots + c_m\mathbf{v}_m \cdot \mathbf{v}_k = 0;$$

since the vectors are orthogonal, we get

$$c_k\mathbf{v}_k \cdot \mathbf{v}_k = 0.$$

Since we know $\mathbf{v}_k \neq \mathbf{0}$, we get $c_k = 0$. This can be done for each $k = 1, 2, \ldots m$ and we see that the set S is linearly independent. □

Definition 12.5.5. An orthogonal set that is a basis is an ORTHOGONAL BASIS. An orthonormal set that is a basis is an ORTHONORMAL BASIS.

Example 12.5.3. The set S from Example 12.5.1 are vectors in $\mathbb{R}^3$. They are orthogonal; thus they are LI. We know $\dim(\mathbb{R}^3) = 3$ thus S is a basis for $\mathbb{R}^3$ and since they are orthogonal, S is an orthogonal basis for $\mathbb{R}^3$. Is S an orthonormal basis? $\|\mathbf{u}_1\| = \sqrt{3}$ so no. (But one could form one from S... remember how?)

Example 12.5.4. The standard bases for $\mathbb{R}^n$, M_{mn}, and P_n are all orthonormal. The vectors in Example 12.5.2 above are another example of an orthonormal basis for $\mathbb{R}^3$.

★ One can always "normalize" an orthogonal basis to get an orthonormal one.

The following theorem hints at one importance of orthonormal bases.

Theorem 12.5.2. Let $\{\mathbf{v}_1, \mathbf{v}_2, \mathbf{v}_n\}$ be an orthonormal basis for a vector space V. Let $\mathbf{u} \in V$. Then $\mathbf{u}$ can be written as a linear combination of the vectors in the basis in the following way:

$$\mathbf{u} = (\mathbf{u} \cdot \mathbf{v}_1)\mathbf{v}_1 + (\mathbf{u} \cdot \mathbf{v}_2)\mathbf{v}_2 + \cdots + (\mathbf{u} \cdot \mathbf{v}_n)\mathbf{v}_n.$$

Proof. We know that we can find coefficients $c_1, c_2, \ldots, c_n$ such that

$$c_1\mathbf{v}_1 + c_2\mathbf{v}_2 + \cdots + c_n\mathbf{v}_n = \mathbf{u}.$$

Now take the dot product of the equation with $\mathbf{v}_k$:

$$\begin{aligned}(c_1\mathbf{v}_1 + c_2\mathbf{v}_2 + \cdots + c_n\mathbf{v}_n) &= \mathbf{u}\cdot\mathbf{v}_k,\\ &\cdots\\ c_k\mathbf{v}_k\cdot\mathbf{v}_k &= \mathbf{u}\cdot\mathbf{v}_k,\end{aligned}$$

since the set is orthonormal, $\mathbf{v}_k\cdot\mathbf{v}_k = 1$ so we have

$$c_k = \mathbf{u}\cdot\mathbf{v}_k.$$

□

Example 12.5.5. Using the above orthonormal basis for $\mathbb{R}^3$ from Example 12.5.2, find the coefficients for $(-1, 2, -3)$.

12.5.2 The Gram–Schmidt process

The Gram–Schmidt process is *a way of finding an orthogonal basis from a given basis.*

Theorem 12.5.3. Let W be a p-dimensional subspace of $\mathbb{R}^n$ and let $\{\mathbf{w}_1, \mathbf{w}_2, \ldots, \mathbf{w}_p\}$ be any basis for W. Then $\{\mathbf{u}_1, \mathbf{u}_2, \ldots, \mathbf{u}_p\}$ is an orthogonal basis for W where

$$\begin{aligned}\mathbf{u}_1 &= \mathbf{w}_1,\\ \mathbf{u}_2 &= \mathbf{w}_2 - \frac{\mathbf{u}_1\cdot\mathbf{w}_2}{\|\mathbf{u}_1\|^2}\mathbf{u}_1,\\ \mathbf{u}_3 &= \mathbf{w}_3 - \frac{\mathbf{u}_1\cdot\mathbf{w}_3}{\|\mathbf{u}_1\|^2}\mathbf{u}_1 - \frac{\mathbf{u}_2\cdot\mathbf{w}_3}{\|\mathbf{u}_2\|^2}\mathbf{u}_2,\\ &\vdots\\ \mathbf{u}_i &= \mathbf{w}_i - \sum_{k=1}^{i-1}\frac{\mathbf{u}_k\cdot\mathbf{w}_i}{\|\mathbf{u}_k\|^2}\mathbf{u}_k.\end{aligned}$$

Proof. The proof is in linear algebra books—read it. Basically the numbers in front of the $\mathbf{u}_k$ render each vector orthogonal to the previous ones. □

Example 12.5.6. $\mathbf{w}_1 = \begin{bmatrix} 1 \\ 0 \\ 1 \\ 2 \end{bmatrix}, \quad \mathbf{w}_2 = \begin{bmatrix} 2 \\ 1 \\ 0 \\ 2 \end{bmatrix}, \quad \mathbf{w}_3 = \begin{bmatrix} 1 \\ -1 \\ 0 \\ 1 \end{bmatrix},$

$$\mathbf{u}_1 = \mathbf{w}_1 = \begin{bmatrix} 1 \\ 0 \\ 1 \\ 2 \end{bmatrix},$$

$$\frac{\mathbf{u}_1 \cdot \mathbf{w}_2}{\|\mathbf{u}_1\|^2} = \frac{2+0+0+4}{1+1+4} = \frac{6}{6} = 1,$$

$$\mathbf{u}_2 = \mathbf{w}_2 - (1)\mathbf{u}_1 = \begin{bmatrix} 2 \\ 1 \\ 0 \\ 2 \end{bmatrix} - \begin{bmatrix} 1 \\ 0 \\ 1 \\ 2 \end{bmatrix} = \begin{bmatrix} 1 \\ 1 \\ -1 \\ 0 \end{bmatrix},$$

$$\frac{\mathbf{u}_1 \cdot \mathbf{w}_3}{\|\mathbf{u}_1\|^2} = \frac{1+0+0+2}{6} = \frac{1}{2},$$

$$\frac{\mathbf{u}_2 \cdot \mathbf{w}_3}{\|\mathbf{u}_2\|^2} = \frac{1-1+0+0}{3} = 0,$$

$$\mathbf{u}_3 = \mathbf{w}_3 - \frac{1}{2}\mathbf{u}_1 - 0\mathbf{u}_2 = \begin{bmatrix} 1 \\ -1 \\ 0 \\ 1 \end{bmatrix} - \begin{bmatrix} \frac{1}{2} \\ 0 \\ \frac{1}{2} \\ 1 \end{bmatrix}$$

$$= \begin{bmatrix} \frac{1}{2} \\ -1 \\ -\frac{1}{2} \\ 0 \end{bmatrix},$$

$$\boxed{\left\{ \begin{bmatrix} 1 \\ 0 \\ 1 \\ 2 \end{bmatrix}, \begin{bmatrix} 1 \\ 1 \\ -1 \\ 0 \end{bmatrix}, \begin{bmatrix} \frac{1}{2} \\ -1 \\ -\frac{1}{2} \\ 0 \end{bmatrix} \right\}}.$$

What if one wanted an orthonormal basis? $\|\mathbf{u}_1\| = \sqrt{6}$, $\|\mathbf{u}_2\| = \sqrt{3}$, $\|\mathbf{u}_3\| = \frac{\sqrt{6}}{2}$ ...

Why do we want orthogonal or orthonormal bases?
Suppose $B = \{\mathbf{b}_1, \mathbf{b}_2, \ldots, \mathbf{b}_p\}$ is a basis for a subspace W. Then, for any $\mathbf{x} \in W$,

$$\mathbf{x} = a_1\mathbf{b}_1 + a_2\mathbf{b}_2 + \cdots + a_p\mathbf{b}_p$$

where these a_i are unique. These a_i are called the COORDINATES of $\mathbf{x}$ with respect to the basis B.

Example 12.5.7. $B = \left\{ \begin{bmatrix} 2 \\ 1 \\ -1 \end{bmatrix}, \begin{bmatrix} 1 \\ 2 \\ 0 \end{bmatrix}, \begin{bmatrix} 1 \\ -1 \\ 1 \end{bmatrix} \right\}$ is a basis for $\mathbb{R}^3$ but not an orthogonal basis.

Express $\mathbf{x} = \begin{bmatrix} 3 \\ 2 \\ -1 \end{bmatrix}$ in terms of B. We would have to solve the system $a\mathbf{b}_1 + b\mathbf{b}_2 + c\mathbf{b}_3 = \mathbf{x}$:

$$\begin{bmatrix} \mathbf{b}_1 & \mathbf{b}_2 & \mathbf{b}_3 & \mathbf{x} \end{bmatrix} \xrightarrow{rref} a = \frac{7}{6}, \quad b = \frac{1}{2}, \quad c = \frac{1}{6}.$$

But if $B_2 = \{\mathbf{b}_1, \mathbf{b}_2, \dots, \mathbf{b}_p\}$ was an orthogonal basis, it would be easier to solve for the coordinates:

$$\begin{aligned} \mathbf{x} &= a_1\mathbf{b}_1 + a_2\mathbf{b}_2 + \cdots + a_p\mathbf{b}_p, \\ \mathbf{b}_i^T\mathbf{x} &= \cdots = a_1\mathbf{b}_i^T\mathbf{b}_1 + a_2\mathbf{b}_i^T\mathbf{b}_2 + \cdots + a_i\mathbf{b}_i^T\mathbf{b}_i + \cdots + a_p\mathbf{b}_i^T\mathbf{b}_p, \\ & a_3 \cdot 0, \\ \Longrightarrow & a_i\mathbf{b}_i^T\mathbf{b}_i = a_i\|\mathbf{b}_i\|^2 = \mathbf{b}_i^T\mathbf{x} \\ \Longrightarrow & a_i = \frac{\mathbf{b}_i^T\mathbf{x}}{\|\mathbf{b}_i\|^2} = \frac{\mathbf{b}_i \cdot \mathbf{x}}{\|\mathbf{b}_i\|^2}. \end{aligned}$$

Thus if a basis is orthonormal, the coefficients with respect to that basis are

$$a_i = \mathbf{b}_i^T\mathbf{x} = \mathbf{b}_i \cdot \mathbf{x}.$$

Example 12.5.8. $\mathbf{x} = \begin{bmatrix} -1 \\ 1 \\ 2 \end{bmatrix}, \quad B = \left\{ \begin{bmatrix} 1 \\ 0 \\ 1 \end{bmatrix}, \begin{bmatrix} -1 \\ 0 \\ 1 \end{bmatrix}, \begin{bmatrix} 0 \\ 1 \\ 0 \end{bmatrix} \right\}$; B is an orthogonal basis.

Write $\mathbf{x}$ in terms of B,

$$\begin{aligned} a_1 &= \frac{\mathbf{b}_1 \cdot \mathbf{x}}{\|\mathbf{b}_1\|^2} = \frac{-1+2}{2} = \frac{1}{2}, \\ a_2 &= \frac{\mathbf{b}_2 \cdot \mathbf{x}}{\|\mathbf{b}_2\|^2} = \frac{1+2}{2} = \frac{3}{2}, \\ a_3 &= \frac{\mathbf{b}_2 \cdot \mathbf{x}}{\|\mathbf{b}_3\|^2} = \frac{1}{1} = 1, \end{aligned}$$

$$\mathbf{x} = \frac{1}{2}\mathbf{b}_1 + \frac{3}{2}\mathbf{b}_2 + \mathbf{b}_3.$$

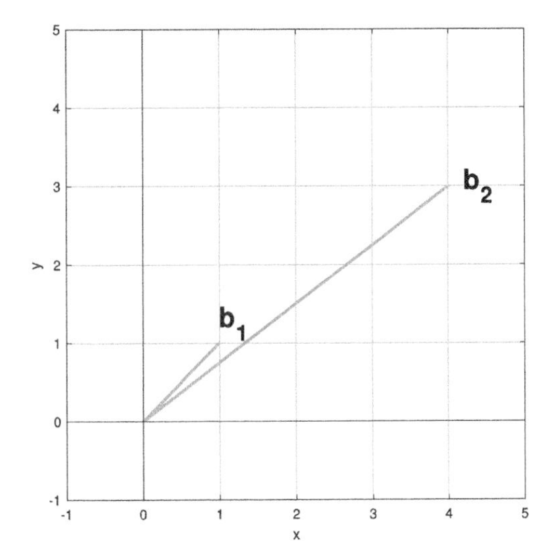

Figure 12.1 A basis for $\mathbb{R}^2$.

Besides finding "coordinates" more easily, there are many other reasons why one would want orthogonal or orthonormal bases for spaces and subspaces.

Example 12.5.9. In this example we will visualize the difference between bases in $\mathbb{R}^2$. Let

$$\mathbf{b}_1 = \begin{bmatrix} 1 \\ 1 \end{bmatrix} \quad \text{and } \mathbf{b}_2 = \begin{bmatrix} 4 \\ 3 \end{bmatrix}.$$

These vectors are linearly independent, so they form a basis for $\mathbb{R}^2$.

If we graph these two vectors, we clearly see that the vectors are not orthogonal and are of different lengths (see Fig. 12.1).

Using the Gram–Schmidt process, we get an orthogonal basis of

$$\mathbf{u}_1 = \begin{bmatrix} 1 \\ 1 \end{bmatrix} \text{ and } \mathbf{u}_2 = \begin{bmatrix} \frac{1}{2} \\ -\frac{1}{2} \end{bmatrix}.$$

If we graph these vectors, we clearly see these are now orthogonal (see Fig. 12.2).

If we normalize these vectors, we get the following orthonormal basis for $\mathbb{R}^2$:

$$\mathbf{v}_1 = \begin{bmatrix} \frac{1}{\sqrt{2}} \\ \frac{1}{\sqrt{2}} \end{bmatrix} \text{ and } \mathbf{v}_2 = \begin{bmatrix} \frac{1}{\sqrt{2}} \\ -\frac{1}{\sqrt{2}} \end{bmatrix}.$$

Graphing these vectors along with the unit circle we clearly see these vectors are not only orthogonal but also of length 1 (see Fig. 12.3).

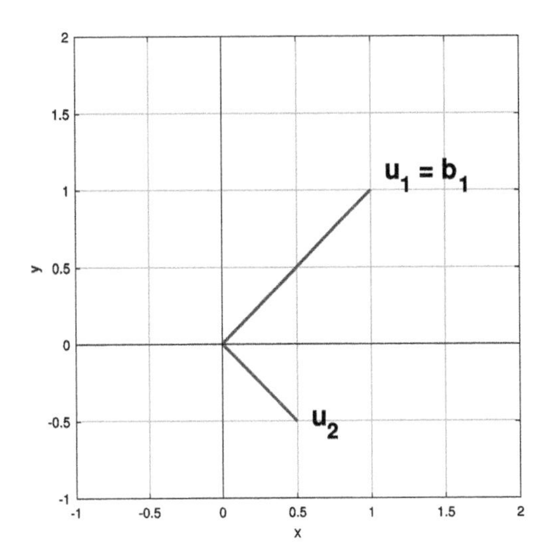

Figure 12.2 Orthogonal basis vectors for $\mathbb{R}^2$.

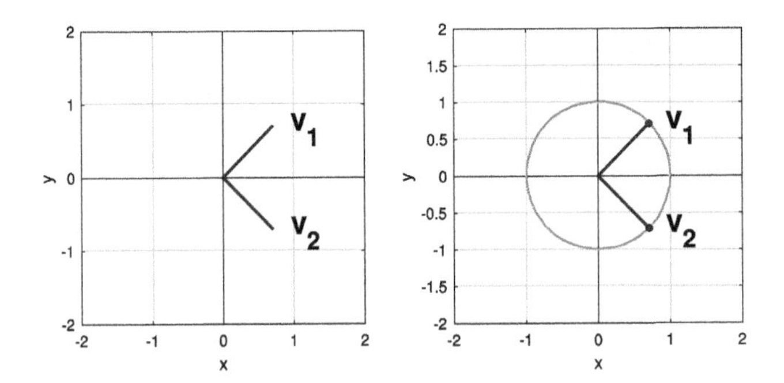

Figure 12.3 Orthonormal basis vectors for $\mathbb{R}^2$.

Example 12.5.10. In this example we will visualize the difference between bases for $\mathbb{R}^3$. Let

$$\mathbf{b}_1 = \begin{bmatrix} 4 \\ 1 \\ -1 \end{bmatrix}, \quad \mathbf{b}_2 = \begin{bmatrix} 1 \\ 1 \\ 0 \end{bmatrix}, \text{ and } \mathbf{b}_3 = \begin{bmatrix} 1 \\ -1 \\ 1 \end{bmatrix}.$$

These vectors are linearly independent, so they form a basis for $\mathbb{R}^3$.

If we graph these three vectors and adjust the view, we see that the vectors are not orthogonal to each other, and they are of different lengths (see Fig. 12.4).

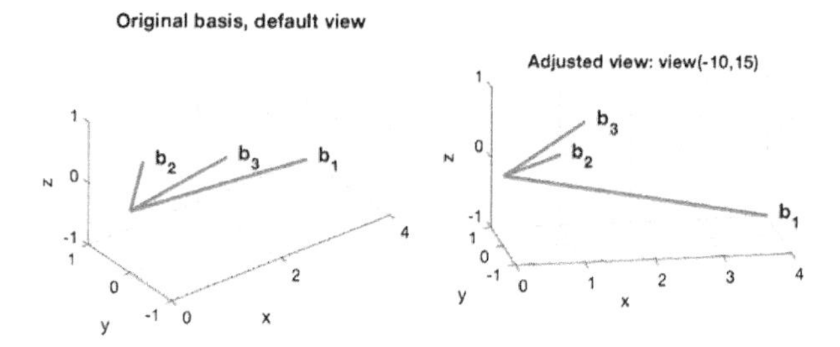

Figure 12.4 Basis vectors for $\mathbb{R}^3$.

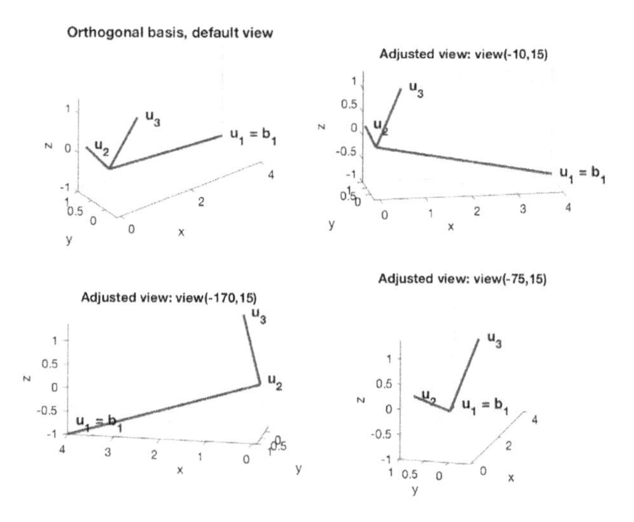

Figure 12.5 Orthogonal basis vectors.

Using the Gram–Schmidt process, we get an orthogonal basis of

$$\mathbf{u}_1 = \begin{bmatrix} 4 \\ 1 \\ -1 \end{bmatrix}, \quad \mathbf{u}_2 = \begin{bmatrix} -\frac{1}{9} \\ \frac{13}{18} \\ \frac{5}{18} \end{bmatrix}, \quad \mathbf{u}_3 = \begin{bmatrix} \frac{5}{11} \\ -\frac{5}{11} \\ \frac{15}{11} \end{bmatrix}.$$

If we graph these and change the view, we can see these are orthogonal vectors (see Fig. 12.5).

If we take these orthogonal basis vectors and normalize them, we get the following orthonormal basis. The unit sphere is also shown so we can see that the vectors are indeed of length 1 (see Fig. 12.6).

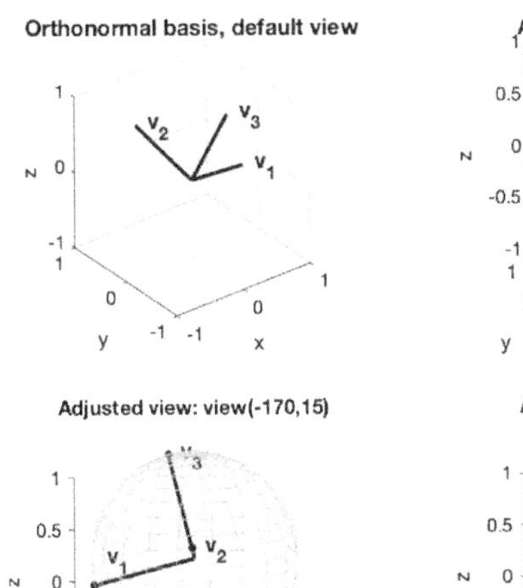

Figure 12.6 Orthonormal basis vectors.

12.6. Answers to example problems

Example 12.1.1

(a) Yes **(b)** No **(c)** No **(d)** Yes

Example 12.1.2

(a) Yes **(b)** No **(c)** Yes **(d)** No **(e)** Yes

Example 12.2.1

(a) Yes. $\mathbf{x} = \mathbf{v}_1 - 2\mathbf{v}_2 + 4\mathbf{v}_3$.
(b) Yes. $\mathbf{x} = 3\mathbf{v}_1 + 0\mathbf{v}_2 - 5\mathbf{v}_3$.
(c) No.

Example 12.2.4

(a) Yes. **(b)** No.

Example 12.2.5 (Answers may vary)

1. $f_1(x) = 3x^2 + 3x - 2$,
2. $f_2(x) = 3x^2 - x - 4$,
3. $f_3(x) = 6x^2 - 7$.

Example 12.3.1

(a) LI **(b)** LI **(c)** LI **(d)** LI **(e)** LD

Example 12.3.7

(a) $\dim(M_{44}) = 16$, **(b)** $\dim(V) = 4$, **(c)** $\dim(W) = 10$.

Example 12.5.5 $\begin{bmatrix} -1 \\ 2 \\ -3 \end{bmatrix} = \frac{-3}{\sqrt{2}}\mathbf{v}_1 + \frac{1}{\sqrt{2}}\mathbf{v}_2 + 3\mathbf{v}_3$, thus the coefficients for this vector with respect to the basis in Example 12.5.2 are $\left(-\frac{3}{\sqrt{2}}, \frac{1}{\sqrt{2}}, 3\right)$.

12.7. Exercises

Note that all ERROR CHECKS within your functions should use the `error` command and display a meaningful error message.

1. Finish the function `gs.m` that is found on the text website. It will have as input a $m \times n$ matrix in which the columns of the matrix are linearly independent; thus the columns of the matrix form a basis for the column space $W \subseteq \mathbb{R}^m$. (We will not check for this linearly independence—we will assume the user has figured this out correctly.) The function will return a matrix in which the columns form an *orthogonal basis* for the column space W. The function will use the Gram–Schmidt process to create the vectors (output matrix).
2. Finish the function `gsON.m` that is found on the text website. It will take as input a $m \times n$ matrix in which the columns of the matrix are linearly independent as in the above problem. (Again, we will not check for this linearly independence—we will assume the user has figured this out correctly.) The function will return a matrix in which the columns form an *orthonormal basis* for the column space W. This function may use the function `gs.m`.
3. For this problem, you will show an example of a basis for $\mathbb{R}^2$ using the Gram–Schmidt process.
   (a) Use the `randi` command to generate a 2×2 matrix with integers from -100 to 100. If the determinant of this matrix equals 0, generate another one until you get a matrix with nonzero determinant (the use of `while` may be useful here). The columns of this matrix form a basis for $\mathbb{R}^2$ and will be the input matrix for `gs` and `gsON`. Make the matrix used output to the command window.
   (b) Use your `gs` function to generate an orthogonal basis from this basis.
   (c) Use your `gsON` function to generate an orthonormal basis from this basis.

(d) Using subplots, generate a 2×2 matrix of plots of these vectors using the `plotVec` function from Exercise 4 in Chapter 6. The first graph will have your original 2 column vectors plotted. The second graph will have the orthogonal basis vectors plotted, and the third graph will have the orthonormal basis vectors plotted. The fourth graph will have the orthonormal basis vectors plotted along with the unit circle (in blue).

For the first two graphs the range on the axes should be from -100 to 100, and the third and fourth graphs from -1 to 1. The first graph the vectors should be in blue, the second in red and the third/fourth in black or another color of your choosing. Make the LineWidth 2 on all of them, have the `grid on` and use the `axis equal` command in all of the subplots.

4. For this problem, you will show an example of a basis for $\mathbb{R}^3$ using the Gram–Schmidt process.

(a) Use the `randi` command to generate a 3×3 matrix with integers from -100 to 100. If the determinant of this matrix equals 0, generate another one until you get a matrix with nonzero determinant (the use of `while` may be useful here). The columns of this matrix form a basis for $\mathbb{R}^3$and will be the input matrix for `gs` and `gsON`. Have the matrix used output to the command window.

(b) Use your `gs` function to generate an orthogonal basis from this basis.

(c) Use your `gsON` function to generate an orthonormal basis from this basis.

(d) Using subplots, generate a 2×2 matrix of plots of these vectors using the `plotVec` function from Exercise 4 in Chapter 6. The first graph will have your original 3 column vectors plotted. The second graph will have the orthogonal basis vectors plotted, and the third and fourth graphs will have the orthonormal basis vectors plotted. The fourth graph will also have the unit sphere plotted using the commands `sphere`, `mesh` and specify a color for the `EdgeColor`. Also, in order to see the vectors inside the sphere, have the command `hidden off` after the `mesh` command and you may want to set the `alpha` value JUST FOR THE MESH COMMAND FOR THE SPHERE set to a number close to 0 (transparent).

For the first two graphs the range on the axes should be from -100 to 100, and the third from -1 to 1. The first graph the vectors should be in blue, the second in red and the third/fourth in black. Make the LineWidth 2 on all of them, have the grid on and use the `axis equal` command in all of the subplots.

5. For this problem we will generate two non-colinear vectors in 3D, plot the plane containing these vectors, the vectors, and the orthogonal and orthonormal basis vectors for this plane. Because of the `subplot` above, you may want to start off this section of code with `clf`. Generate a random 3×2 matrix with values between

−2 and 2 (integers or real numbers—your choice). If the cross product of the two columns equals the zero column vector, keep generating a new random matrix until the cross product is nonzero. Plot the plane that contains these two vectors using `plotPlane` (from Exercise 5 in Chapter 6), with domain from −2 to 2, and `EdgeColor` a gray color. Using `gs` and `gsON`, find the orthogonal basis vectors and orthonormal basis vectors for this same plane. Plot the original vectors in a blue color, the orthogonal vectors in a reddish color and dotted, and the orthonormal vectors in black. Experiment with the `LineWidth` so you can see the vectors.

NOTE: For all of the above 3D graphs, you may use the `view` command so you can see all of the vectors and you compare between the original and orthogonal, but each time you rerun it the best view needed may be different because of the different random vectors generated. Do not worry if the 3D view on any published page or turned in work is not ideal.

APPENDIX A

Publishing and Live Scripts

A.1. Live scripts

New features of Live Scripts are being added with almost every new update of MATLAB®. Thus this will not be a detailed explanation of Live Scripts.

Live Scripts have extensions ".mlx" and as mentioned above, can only be viewed and edited within MATLAB. One can create a Live Script in several ways. You can convert an already written script to a Live Script by saving the file and selecting the format "MATLAB Live Code Files (.mlx)." You can also start a new Live Script by selecting "New > Live Script" within the Editor tab.

A nice feature of Live Scripts is that the code and output or figure are displayed side-by-side by default, while a published document has a vertical layout. Thus one can easily modify the code and see the result. Another wonderful feature of Live Scripts is the ability to add "Controls" such as sliders, drop down menus, etc. The Live Editor has the ability to have code interspersed with formatted text, similar in style to a published script file. One can Export the Live Script to several formats, similar to the options of publishing script files and the look of the resulting document is also similar. One should note that when you export a Live Script with controls, what is shown is the currently selected control. See the text website for examples of Live Scripts.

A.2. Basic scripts or M-files

The m-files and the published pages for these examples can be found at on the text website.

Here I will explain some of the syntax needed to publish m-files to webpages (HTML files). If you are doing this for an assignment or project, the specific directions of where to save these files, etc. are given separately.

First, you must understand the difference between a BASIC SCRIPT FILE (m-file) and a PUBLISHABLE SCRIPT FILE. A script file is just an ASCII (American Standard Code for Information Interchange, i.e., basic text) file with an extension of `.m`. Within that file are MATLAB commands as if you had typed them in at the command prompt within the Command Window. You can put comments (and should, especially for long files) to better read/debug the file. Comments start with "%". Here is an example:

Link 1: basicMFile.m

```
% Example of Basic Script file
```

```
% Lisa Oberbroeckling, 2019
clc
x=linspace(-pi,pi);
y=sin(x);
plot(x,y)
% next problem
A=[1 2 3;4 5 6;7 8 9];
B=[A(1,:); -4*A(1,:) + A(2,:); A(3,:)]
B2=[B(1,:); B(2,:); -7*B(1,:) + B(3,:)]
C=[B2(1,:); -1/3*B2(2,:); B2(3,:)]
```

A.3. Publishing M-files

There are several different ways you can publish your m-files to HTML files using MATLAB. One way is to enter the following on the command line within the Command Window: `publish('filename.m')`.

You can also type: `publish('filename.m', 'html')`. You would do this if you have changed the publishing settings to be "latex", for example.

The other way of publishing, which is more common, is to do it within MATLAB's Editor Window and press the "Publish" button (or "Save and Publish" button if changes have been made without saving).

This will create a folder named "html" in the current working directory (if there is not a folder of that name already there), and put the `filename.html` file and any other files necessary for the webpage (PNG files for images, for instance) in the "html" folder. If we publish the above file, it would not be very pretty. See Link 1 on the course website.

Notice that the output that is shown on the webpage is out of order of the commands given; the output of the next problem appears BEFORE the plot. Thus we want to format the comments in a special way so that when it is published, the MATLAB commands and output appear in order. This is the topic of the next section.

A.4. Using sections

In order to format our m-file to make it better for publishing, we want to break up the commands in the m-file into SECTIONS. Each SECTION is broken up by "`%%`". When you do this within the Editor Window, you will notice lines appearing between each section. Sections are useful not only for publishing, but for running and debugging scripts. Sections are used in publishing to signify different sections of the page (like for homework assignments to have different sections for each problem). Sections also determine how/where output for lines of code are displayed. For more detailed information, see the MATLAB Documentation on Publish and Share MATLAB Code [17].

A.4.1 Using sections for publishing

If you have text following the "%%" on the same line, this also creates a SECTION with that text as the section title. In addition, a bulleted list is created with those linked section titles at the top of the webpage. For the following we just added "%%" to two lines (lines 1 and 7); compare with basicMFile.m on page 215.

Link 2: publishMFile1.m

```
%% Example of Basic Script file
% Lisa Oberbroeckling,  2019
clc
x=linspace(-pi,pi);
y=sin(x);
plot(x,y)
%% second problem
A=[1 2 3;4 5 6;7 8 9];
B=[A(1,:); -4*A(1,:) + A(2,:); A(3,:)]
B2=[B(1,:); B(2,:); -7*B(1,:) + B(3,:)]
C=[B2(1,:); -1/3*B2(2,:); B2(3,:)]
```

The above is better than the published page without sections, but can be better. We may want the page to start with a title. This is done by adding a line "%%" after our title (and another comment line(s) for other introductory text, like my name).

LINK 3: publishMFile2.m

```
%% Example of Basic Script file
% Lisa Oberbroeckling, 2019
%%
clc
x=linspace(-pi,pi);
y=sin(x);
plot(x,y)
%% second problem
A=[1 2 3;4 5 6;7 8 9];
B=[A(1,:); -4*A(1,:) + A(2,:); A(3,:)]
B2=[B(1,:); B(2,:); -7*B(1,:) + B(3,:)]
C=[B2(1,:); -1/3*B2(2,:); B2(3,:)]
```

If you look at the published webpage, you will notice that we have a section link and title for the "second problem" but not for the first. So we probably want to change line 3 to include a section title:

LINK 4: publishMFile3.m

```
%% Example of Basic Script file
% Lisa Oberbroeckling, 2019
%% first problem
clc
x=linspace(-pi,pi);
y=sin(x);
plot(x,y)
%% second problem
A=[1 2 3;4 5 6;7 8 9];
B=[A(1,:); -4*A(1,:) + A(2,:); A(3,:)]
B2=[B(1,:); B(2,:); -7*B(1,:) + B(3,:)]
C=[B2(1,:); -1/3*B2(2,:); B2(3,:)]
```

When publishing m-files, each time a new section is started, MATLAB displays the output created by the commands of the previous section. The difference between `publishMFile3.html` and `publishMFile4.html` is where the output is displayed for the second problem.

LINK 5: publishMFile4.m (partial view)

```
%% second problem
% problem 2a
A=[1 2 3;4 5 6;7 8 9];
B=[A(1,:); -4*A(1,:) + A(2,:); A(3,:)]
%%%
% problem 2b
B2=[B(1,:); B(2,:); -7*B(1,:) + B(3,:)]
%%
% problem 2c
C=[B2(1,:); -1/3*B2(2,:); B2(3,:)]
%%
% problem 2d
x=linspace(-10,10);
y=exp(x);                    % can comment after a command, too
plot(x,y)
title('Another Example')
```

Another important place to insert a section break is when you want to have text following MATLAB commands, but within the same section. If you just include comments after the MATLAB commands, they will be formatted as comments within the displayed code, not as text. Instead, insert a section break (without a section title) and then the comment block that will be the text:

LINK 5: publishMFile4.m (partial view)

```
%% third problem
A=[1 2 3;4 5 6;7 8 9;eye(3)]
% If I don't have a cell break above this comment, this
% text just appears as comments within the command lines,
% and not text on the webpage.
%%
% Instead, have a cell break or a section break
```

Note that you can also have section titles without having section breaks. This is done by having the line start with `%%%`" along with the section text. This will have the text and/or MATLAB commands be within the sections, but the output of those commands at the next section break, which may not have the desired effect.

LINK 5: publishMFile4.m (partial view)

```
%%% Next section
% this section does not have a cell break.  This may
% or may not be useful depending on how you want the
% output displayed on the published webpage. It works
% here because this is the last section and cell.
x=linspace(0.0001,10);
y=log(x);
plot(x,y)
```

Using sections is especially important for m-files with multiple plots. Remember, MATLAB only shows the last plotting command (like `plot`, ..., `mesh`, `surf`, etc.). You can have multiple commands appear on the same figure by using the `hold on` and `hold off` commands. But if you want to display multiple figures (not in the same window), you have to either use the `pause` command or the `figure` command. This first example uses the `pause` command:

LINK 6: publishMFile5.m

```
%% Example of Basic Script with pause
% Lisa Oberbroeckling, 2019
%%
x=linspace(-pi,pi);
y=sin(x);
plot(x,y)
hold on
y=cos(x);
plot(x,y,'r')
hold off
```

```
title('First Plot')
pause
[x,y]=meshgrid(linspace(-10,10));
z=sin(x).*cos(y);
mesh(x,y,z)
xlabel('x'),ylabel('y'),zlabel('z')
title('Second Plot')
```

If you run the file publishMFile5.m, the first figure will appear and then MATLAB will be paused. The second will appear after any key is pressed. When this is published, even the publishing will be on pause after the first figure is created until you press a key. But if you look at the webpage, only the last figure is actually shown on the webpage. As discussed above, when publishing, MATLAB runs each section as a block and then displays any output. At the end of the section the only output that MATLAB sees as being created by the section of commands is the second figure. The second figure replaces the first figure, so it is not shown on the webpage. Thus, we need to create a section for each figure we want on the webpage. When we publish the m-file, we have to remember to "press any key" for the publishing can continue, which is really annoying so you may want to take the `pause` command out or comment it out. In the following example we created a new section without a section title.

LINK 7: publishMFile5b.m

```
%% Example of Basic Script with pause
% Lisa Oberbroeckling, 2019
%%
close all
clc
% first plot
x=linspace(-pi,pi);
y=sin(x);
plot(x,y)
hold on
y=cos(x);
plot(x,y,'r')
hold off
title('First Plot')
% pause
%%
[x,y]=meshgrid(linspace(-10,10));
z=sin(x).*cos(y);
mesh(x,y,z)
xlabel('x'),ylabel('y'),zlabel('z')
title('Second Plot')
```

The next group of files use the `figure` command.

LINK 8: publishMFile6.m

```
%% Example of Basic Script with figure
% Lisa Oberbroeckling, 2019
%%
close all
clc
% first plot
figure(1)
x=linspace(-pi,pi);
y=sin(x);
hold on
y=cos(x);
plot(x,y,'r')
hold off
plot(x,y, 'r')
title('First Plot')
% Second plot
figure(2)
[x,y]=meshgrid(linspace(-10,10));
z=sin(x).*cos(y);
mesh(x,y,z)
xlabel('x'),ylabel('y'),zlabel('z')
title('Second Plot')
% third plot
figure(3)
[x,y]=meshgrid(linspace(-10,10));
z=x.*cos(y);
mesh(x,y,z)
xlabel('x'),ylabel('y'),zlabel('z')
title('Third Plot')
% fourth plot
figure(4)
[x,y]=meshgrid(linspace(-10,10));
z=x.*y;
mesh(x,y,z)
xlabel('x'),ylabel('y'),zlabel('z')
title('Fourth Plot')
```

When the above file is published, it puts each figure side-by-side on one line. Depending on how many figures you have this may not have the desired effect, so you may want to have each figure within its own section instead.

LINK 9: publishMFile6b.m

```
%% Example of Basic Script with figure
% Lisa Oberbroeckling, 2019
```

```
%%
close all
clc
%% first plot
figure(1)
x=linspace(-pi,pi);
y=sin(x);
hold on
y=cos(x);
plot(x,y,'r')
hold off
plot(x,y, 'r')
title('First Plot')
%% Second plot
figure(2)
[x,y]=meshgrid(linspace(-10,10));
z=sin(x).*cos(y);
mesh(x,y,z)
xlabel('x'),ylabel('y'),zlabel('z')
title('Second Plot')
%% third plot
figure(3)
[x,y]=meshgrid(linspace(-10,10));
z=x.*cos(y);
mesh(x,y,z)
xlabel('x'),ylabel('y'),zlabel('z')
title('Third Plot')
%% fourth plot
figure(4)
[x,y]=meshgrid(linspace(-10,10));
z=x.*y;
mesh(x,y,z)
xlabel('x'),ylabel('y'),zlabel('z')
title('Fourth Plot')
```

Note that you can also use sections for running/debugging code. This topic is only covered briefly.

A.4.2 Using sections for running/debugging files

Using sections is for running portions of your code is especially useful. You can separate out self-contained portions of your code and just run that piece. This is especially useful for long homework assignments within one file; just run on problem at a time to see if runs as expected. You can run the code within a section block several ways.

1. While the cursor is within the section you want to run, use the keystrokes of Ctrl-Enter (Cmd-return on a Mac). Using the keystrokes of Ctrl-Shift-Enter (Cmd-return-enter) will run the current section and put the cursor within the next section.

2. Within the editor, select the "Run Section" or "Run and Advance" buttons on the Editor Tab.

WARNING: when you run (evaluate) a section, the file may not be saved automatically as it is when you run the entire m-file!

A.5. Formatting text

You can format your m-file by clicking on the "Publish" tab. By clicking on one of the items, MATLAB will insert text into your m-file for that purpose. This includes inserting text for things already discussed above, like inserting a section break, title, etc. There are also buttons on the toolbar within the editor window for some of these items. You also have the ability to customize your toolbar to add others.

Going to the menu and/or using the buttons on the toolbar can take extra time after awhile, so it is also useful to know how to just type in the formatting.

A.5.1 Basic text formatting

In order to create a new line or new paragraph of text, have a blank comment line in between the lines.

Link 10: publishMFile7.m (partial view)

```
%
% Text can be *bold*, _italic,_ and/or |monospaced|.
% One can also combine these formats like:
%
% _*BOLD, ITALIC TEXT*_
%
```

Text can be formatted to be **bold**, *italic*, `monospaced`, or combinations such as ***bold and italic***.

Link 10: publishMFile7.m (partial view)

```
%
%% Unordered (Bulleted) List
%
% * first item
% * second item blah blah
%
```

A.5.2 Lists

One can have an unordered, or bulleted list.

Link 10: publishMFile7.m (partial view)

```
% without a section title.  You also must have a blank comment line to end the list.
    Here's another example.
%%
%
% * item number 1
% * item number 2
```

Keep in mind that you must have a section break before the list, with or without a section title. You also must have a blank comment line to end the list. Here is another example.

Link 10: publishMFile7.m (partial view)

```
%
% # first item
% # second item blah blah
%
% As in the bulleted list, one must have a cell break, with or without a
```

You can also have an ordered (numbered list) using the same formatting as above, but with "#" instead of "*" for each list item.

Link 10: publishMFile7.m (partial view)

```
% section title.  Second example without section title:
%%
%
% # blah blah blah blah blah blah blah blah blah blah blah blah blah blah blah blah
    blah blah blah blah
% # yadda yadda yadda yadda yadda yadda yadda yadda yadda yadda yadda yadda yadda
    yadda yadda yadda yadda yadda yadda yadda
```

As in the bulleted list, one must have a section break, with or without a section title. Second example without section title:

Link 10: publishMFile7.m (partial view)

```
% You can have the links display the URL or display other text.
```

```
%
% Here's an example of using the URL as
% the text of the link: <http://www.mathworks.com>
%
```

A.5.3 HTML links

You can have the links display the URL or display other text. URL as the link:

Link 10: publishMFile7.m (partial view)

```
% *LINKING THE M-FILE* It may be nice (required!) to link the M-File that
% was published to create the webpage.
```

You can have any text for the link:

Link 10: publishMFile7.m (partial view)

```
% <../publishMFile7.m M-File for this page>.
%
```

LINKING THE M-FILE: It may be nice (required!) to link the m-file that was published to create the webpage. The easiest way to do it is like it is done below rather than using the entire URL.

Link 10: publishMFile7.m (partial view)

```
% the file.  The editor might automatically wrap the text if the URL
% and/or the text for the link is long.  If this is the case, go back and
% make it one line.
%
```

The "../" before the filename means to go back one folder from where the HTML file is located, which is where the m-file is located.

IMPORTANT: the text within the "<" and ">" must be on the same line within the file. The editor might automatically wrap the text if the URL and/or the text for the link is long. If this is the case, you must go back and make it one line. Otherwise, the link will be broken!

A.5.4 Inserting images

Any figures that the MATLAB code creates will automatically be saved as PNG files and inserted on the webpage. You can also include images like the following example.

Link 10: publishMFile7.m (partial view)

```
% into which the HTML file the M-file produced is located.
%% Preformatted Text
%
```

The image must be on its own line; no text can appear before or after it for the image to be shown correctly on the page. Also note that the above is assuming the file for the image is located in the "html" folder into which the HTML file the M-file produced is located.

A.5.5 Pre-formatted text

Pre-formatted text is a way to display text on the webpage EXACTLY as it appears in the editor, including extra spaces, line breaks in exactly the same place, etc. This is commonly used for displaying lines of code in programming.

IMPORTANT: notice that the code below does not appear any different from plain comments. In order for it to be pre-formatted, you must use the menu item to insert the lines and change it. Here is what the menu item inserts:

Link 10: publishMFile7.m (partial view)

```
%
%  preformatted text
%   displayed
%  exactly
```

Then you change the words "PREFORMATTED" AND "TEXT" and add lines if necessary:

Link 10: publishMFile7.m (partial view)

```
%  ---------------------------------------------------
%  function y = myexample(x)
%  % MYEXAMPLE(X) is a function for example purposed only
%  %
%  y = x.^2;
%  ---------------------------------------------------
%
%
%% Inserting HTML Code
%
% <html>
% <table border="1" cellspacing="0" cellpadding="3"><tr><td>one</td><td>two</td></tr
   ></table>
% </html>
```

A.5.6 Inserting HTML code

You can insert other HTML code (such as tables) into your document.

Link 10: publishMFile7.m (partial view)

```
%
% One can insert basic LaTeX commands and equations.  One can have LaTeX
% code with the paragraph like
% $$ e^{\pi i} + 1 = 0 $$
% or on a separate line.
```

A.5.7 Inserting LaTeX equations

Basic LaTeX equations can be displayed on the webpage. Only basic mathematics can be displayed when publishing to HTML mode, and some symbols do not display correctly. For example, the not equals "$\neq$" symbol did not work in earlier versions, then it worked in version 2012a, and so on. Technically speaking, when these equations are published to an HTML file, each equation is saved as a PNG image (in the "html" folder or same folder the HTML file is saved) and that image file is displayed on the webpage.

Link 10: publishMFile7.m (partial view)

```
%
%
%
% Notice that the "not equals" symbol may or may not work, depending on the version
    of MATLAB!
% $$ 0 \ne 1 $$
%% M-file that created this page
% <../publishMFile7.m publishMFile7.m>
```

These images make the formulas appear blurry and/or small, and as mentioned above, not every symbol will display correctly. Also, it could make the webpage slower to load and would not be as accessible. Thus it may be better to use MathJax if creating a webpage. If there is a need for a lot of LaTeX or more complicated math typesetting, consider publishing it as a LaTeX file.

APPENDIX B

Final Projects

B.1. Ciphers

Create function(s) that create a substitution cipher encoder and decoder and a transposition cipher encoder and decoder, plus functions that demonstrate these ciphers. Some useful MATLAB® functions: `randperm`, `double`, `char`, and `reshape`.

B.1.1 Substitution cipher

The encoder function will create a basic substitution cipher on an "extended alphabet". The encoder function will have two inputs: the first is string (entered in single quotes) that is a filename for a text file in which the text may include numbers, characters, and spaces. The second input will be a flag (either 1 or 0) as to whether the messages are displayed.

You will use the command `fileread(filename)` to open the file and store the text as one long vector of strings. Any characters that have an ASCII code of 32–126 are allowed: see http://www.asciitable.com/. (ASCII stands for "American Standard Code for Information Interchange.") The encoder function will then encode the text using a random substitution cipher. The function will return a string that is the encoded text along with a matrix of characters that will be the key. The top row of the matrix will be the original alphabet (represented by the ASCII code), and the second row will be the substituted alphabet (represented by the ASCII code). Also, if the flag is true for displaying the messages, the encoder function will display both the original and encoded message to the screens. The display will not be as one long string; instead, the messages will be formatted to be split across several lines so that no more than 50 characters are on one line. If the flag is false, no output is displayed to the screen (unless the user chooses to see the output of the function).

The decoder function will have three inputs; the key (which would be the matrix that the encoder function gives you), the filename of the encoded text, and the flag for displaying the messages to the screen. Using the key, the decoder function will return a string of text that is the decoded message. If the flag is true for displaying the messages, then both the encoded message and decoded message are displayed to the screen as in the encoder function. If the flag is false, no output is displayed to the screen (unless the user chooses to see the output of the function).

IMPORTANT: do not choose a specific substitution cipher—have MATLAB randomly choose it so that each time you run the encoding function, you may get a different encoded message (thus the need for the key to decode it).

To demonstrate the above functions, create another function that will create tables that show the substitution cipher. It will as input take the "key" and create tables of the character and what that character looks like in the encoded message. Have one table be the characters (may split up into multiple tables for visual purposes), another table the numbers, and one or two tables the letters (split between upper and lower? you will be the judge). Thus your output *may* look like this (example cipher shown):

```
 A B C D E F G H I J K L M N O P Q R S T U V W X Y Z
----------------------------------------------------
 D E F G H I J K L M N O P Q R S T U V W X Y Z [ \ ]


 a b c d e f g h i j k l m n o p q r s t u v w x y z
----------------------------------------------------
 d e f g h i j k l m n o p q r s t u v w x y z { | }


 0 1 2 3 4 5 6 7 8 9
--------------------
 3 4 5 6 7 8 9 : ; <


   ! " # $ % & ' ( ) * + , - . / : ; < = > ? @ [ \ ] ^ _ ` { | } ~
-------------------------------------------------------------------
 # $ % & ' ( ) * + , - . / 0 1 2 = > ? @ A B C ^ _ ` a b c ~   ! "
```

This would signify that for every "A" in the original message, it is encoded as "D", every space becomes the character "#", etc. Make these tables visually pleasing in the output window (using `fprintf`, etc.).

B.1.2 Columnar transposition cipher

For the columnar transposition cipher, the encoding function has as input filename of the original text (similar to above: characters, spaces, numbers, etc. are allowed), the number of columns to use, and a flag for displaying the messages (similar as above). The number of rows in the "matrix" is determined by the number of characters in the given text; if there are not enough characters to completely fill the matrix, dummy (random) characters are put at the end of the text to complete the matrix. The output will be the encoded text and a key which is a vector that tells the order of the permuted columns. The decoder function will have as input the filename that contains the encoded text, the key (vector), and a flag for displaying the messages. The output is the decoded message.

Another function will demonstrate the columnar transposition. It will take as input the original message and the key. The function will display the original message, the matrix of characters, the permuted matrix of characters, and the encoded message. These displays should be formatted such that if the messages are long, then no more than 50 characters are displayed on one line.

Your report should include a basic history of these types of ciphers, examples, pros and cons to using these types of ciphers, and other variations of the ciphers.

B.2. Game of Pig

You will create a **script file** in which the user will use to play the game of Pig against the computer. The script file will use the `input` command.

Rules of the game

Object: get to 100 points before your opponent.

A player's turn involves rolling a standard six-sided die.

- If anything other than a 1 is rolled, the number rolled is added to the player's subtotal. The player can choose to roll again, or stop their turn and the subtotal is added to the player's total.
- If a 1 is rolled, the player's turn is over and no points are added to the player's total.

A running total of both player's scores are always displayed.

For the user's turn, your script will roll the die and display the number rolled. If the number is 2 through 6, you will ask whether the user wants to roll again and a subtotal is displayed, etc.

For the computer's turn, if the number 2 through 6 is rolled, the computer will have a certain algorithm that will determine whether to roll again. There may be an initial probability p of rolling again or different probabilities under various conditions of total scores and subscores, etc. Also, at each subsequent roll, the computer should be less likely to roll again than before. You can choose your algorithm and value of p and how the probability is reduced with each subsequent roll (as long as it fulfills the requirements). In your final report you will explain how you chose the algorithm and probability value(s). You do not need to get fancy with figuring out the "optimal strategy" to figure out these numbers (that would be material for another class).

At the beginning of each turn, whether it is the user's turn or the computer's, the screen is cleared and the both players' totals are displayed (and continue to be displayed). For each roll of the die, the roll is displayed, and the current subtotal (if applicable). Thus you see each of the computer's rolls within a turn. The values displayed should be clearly labeled and easy to read (use of `fprintf` will be helpful here).

Your report should include some possible variations and possible improvements to your game.

B.3. Linearization and Newton's method

One application of differentiation is "linearization;" using the tangent line to a function to approximate values of the function near the point at which the tangent line is calculated. Finding the linearization of $f(x)$ at $x = a$ is equivalent to finding the tangent line to $f(x)$ at $x = a$. You will create a function to help demonstrate linearization and to use it to approximate values.

B.3.1 Linearization

Create a function called `myLinear` in which the function takes a string that is the function $f(x)$, and a value a. Using the commands `syms`, `str2sym`, `subs`, and `diff`, your function will calculate the tangent line to $y = f(x)$ at the point $x = a$. The function will return the linearization $L(x) = f'(a)(x - a) + f(a)$ (as a symbolic function).

In a separate script file that you may publish, you will use your function to do the following problems.

1. Consider the function $f(x) = \sqrt[3]{1+x}$.
   - **(a)** Use your `myLinear` function to find the linearization of $f(x)$ at the point $x = 0$. Careful! You may have to use a different function than `(1+x)^(1/3)` or `nthroot(1+x,3)` in order to make it work (`surd`?). Display the answer.
   - **(b)** Use `myLinear` function to approximate the values of $\sqrt[3]{0.95}$ and $\sqrt[3]{1.1}$.
   - **(c)** Graph the function $y = f(x)$, the tangent line at $x = 0$, and the points corresponding to the approximations of $\sqrt[3]{0.95}$ and $\sqrt[3]{1.1}$ on the same graph. Create several graphs, zooming in to see the difference between the graph and the tangent line at these points. Make sure you create a legend to make things clearer.
2. Let $f(x) = (x - 1)^2$, $g(x) = e^{-2x}$, and $h(x) = 1 + \ln(1 - 2x)$.
   - **(a)** Use your `myLinear` function to find the linearizations of f, g, and h at $a = 0$. What do you notice? Why did this happen?
   - **(b)** Graph f, g, h, and the tangent lines on one graph. Create a legend in order to tell which is which. For which function is the linearization a better approximation (and for approximately what x-values)? Create several graphs, zooming in, to support your answers.

B.3.2 Newton's method

Another application of derivatives is Newton's method for finding roots.

Create a function `newton` that uses Newton's method to find an approximate solution to the equation $f(x) = 0$, for a given function f. The input should be:

- the function f (given as a string with x as a variable),
- an initial guess x_0 for the solution,

- the desired accuracy (ERROR CHECK: this number should be positive; if not, an appropriate error message should be displayed using the `error` command),
- and the maximum number of iterations allowed (so it will stop if accuracy cannot be reached) (ERROR CHECK: this number should be a positive integer; if not, an appropriate error message should be displayed using the `error` command).

The function uses the Symbolic Math Toolbox (`syms`, `diff`, etc.) to convert the given string for the function to a symbolic function and to calculate the derivative of the given function f. Your HELP lines should make clear what the inputs are and the order. The output will be the approximation to the solution of $f(x) = 0$. Your code will iterate until the absolute value of the difference between the last two iterations is less than the desired tolerance/accuracy OR the maximum number of iterations has been reached. In either case, **an appropriate message should be printed on the screen** so the user knows if desired accuracy has been reached or not. The message should include how many iterations were completed. Use either `disp` or `fprintf`, and/or `sprintf` for this message; experiment with this. The output of your function is the LAST iterate. In other words, if x_7 was calculated to determine that x_6 is accurate enough, still output x_7 and state that 7 iterations were calculated. If the derivative ever equals 0 at any iteration, an appropriate error message should be displayed and the function will stop (use the `error` command).

The following problems will demonstrate/use Newton's method. The code for these problems will be in the same script file as the problems on Linearization.

3. Consider the function $f(x) = x^4 - x - 1$.
   (a) Use your `newton` function to find x_2 using $x_1 = 1$ to find an approximation to a root of $f(x)$.
   (b) Graph $y = f(x)$, the tangent line at $x_1 = 1$, and the tangent line at x_2 to see how the roots of each subsequent tangent line gets closer to the root. Make sure you have a legend and appropriate axes to be able to see everything.
4. Use your function `newton` to find all roots to the equation $e^{\arctan(x)} = \sqrt{x^3 + 1}$ correct to eight decimal places (make sure all decimal places are displayed). In order to do this, first create an appropriate plot to figure out what you are going to use for your initial approximations of x_1 for each root.
5. To demonstrate the importance of that first guess, consider the function $f(x) = x^3 - x - 1$.
   (a) Use your function `newton` to find a root of the equation correct to six decimal places using an initial approximation of $x_1 = 1/\sqrt{3}$. State how many iterations were needed or why Newton's method did not work in this case.
   (b) Use your function `newton` to find a root of the equation correct to six decimal places using an initial approximation of $x_1 = 0.57$. State how many iterations were needed or why Newton's method did not work in this case.

(c) Use your function `newton` to find a root of the equation correct to six decimal places using an initial approximation of $x_1 = 0.6$. State how many iterations were needed or why Newton's method did not work in this case.

(d) Use your function `newton` to find a root of the equation correct to six decimal places using an initial approximation of $x_1 = 1$. State how many iterations were needed or why Newton's method did not work in this case.

(e) Graph the function $y = f(x)$, and the tangent lines at $x = 1/\sqrt{3}$, $x = 0.57$, $x = 0.6$, and $x = 1$. Does the graph explain your answers to the above? Make sure your graph has a legend and appropriate axes (or create a second graph with appropriate axes) to see what is going on and support your claims.

NOTE: if any of the above return an error, you may need to comment out that code so the rest of your script can run afterwards (keep the code in there to "show your work" and mention something in the comments/text of your report).

B.4. Disk and Shell method

Create demonstrations of the Disk and Shell methods for volumes of revolution.

For the disk method, we will consider the solid of revolution by revolving the region formed by $y = f(x)$, $y = 0$, $x = a$, and $x = b$ about the line $y = k$. For the shell method, we will consider the solid of revolution by revolving the same region about the line $x = k$. To make it easier, we will assume that the lines of revolution lie completely outside or on the border of the revolved region.

Your function(s) will have as inputs the function $f(x)$ (entered as a string), a, b, k and the positive integer n. Your function should check that $a < b$ and that either x or y is given and that n is a positive integer. Your function(s) will do the following:

- Create a figure that shows the region and the line of revolution with the region shaded/colored and an appropriate title.
- Form n rectangles of width $\Delta x = (b - a)/n$ and height of each rectangle is $f(x_k^*)$, where x_k^* is the midpoint of the kth subinterval. These rectangles along with the original region will be graphed in a second figure.
- Revolve these rectangles to form the approximating disks (or approximating shells), and plot these disks (shells) in 3D. You will also calculate the volume of these approximating disks (shells) and display the answer both in the title of the plot and as output of the function.
- Show the volume of revolution in 3D.

In all of the figures created, axes should be labeled, appropriate titles created, etc. They should not appear jagged so appropriate domains/views will need to be defined.

Your report should show the use of these for various functions and/or values of n, and different lines of revolution.

Some useful MATLAB commands are `sphere` and `cylinder` discussed in Section 4.5.1.

B.5. Power ball data

Practice exploring data with Powerball numbers. There are many things that can be explored about the Powerball numbers that have been drawn. The rules of the Powerball game have changed, and thus any of the statistics and graphs should reflect these changes. An explanation of the these changes should be included in the report. Visualization of the data is preferred and should be clear with titles, legends, etc. Many times the way the data are given to us need to be cleaned up and modified before it is in a usable format. Anything that needs to be done to accomplish this should be documented and explained in the report.

Use the PowerballNumbers.txt data file to load the Powerball numbers drawn with their dates. Then explore away! A common numerical summary of the columns (numbers drawn). Frequency, histograms, min, max, average, median, mode, etc. These may be done based on the white ball order and overall. What are the differences or gaps between the numbers in any drawing (average, min, max)? Is there anything interesting when you look at the sums of the numbers drawn? Were any winning drawings repeated? What about the white ball only combo? If so, how many? Were any drawings in which the Powerball was the same number as one of the white balls? If so, how many times? Is there any number not been selected? Have there been any consecutive numbers drawn? If so, how many? Any time three consective numbers drawn? Etc.

Can you generate your own random Powerball tickets? How many times does it take to generate a drawing that has already been a winner? How about you pick a specific drawing date (like near your birthday...) and see how long it may take to generate that winning drawing?

Use the PBJackpots.xlsx file to load the dates of winning jackpots and their amounts. Matching the dates with the other data, is there anything different about the numbers that end up being jackpots and the numbers that don't? What about the summaries about the size of the jackpots? The last rule change was supposed to achieve higher jackpots. Does this appear to be true? If you graph the size of the jackpots over time, is there a trend on spikes, etc.?

APPENDIX C

Linear Algebra Projects

C.1. Matrix calculations and linear systems

C.1.1 First handout

The following is the basic information given to the students for the introductory project followed by problems that have been a part of this project. Additional information of how the project is saved and submitted is omitted. The submissions can be given in the form of a `diary` of the commands, a script file, published script file (to a webpage, PDF, etc.) or a Live Script.

Entering a matrix

Rows of a matrix are separated by semicolons (;) and entries within a row are separated by either commas or spaces. Brackets enclose the matrix. For example, you could use either of the following commands to define the same matrix.

```
A=[1,2,-1,2,0;2,1,1,-1,0;3,-1,-2,3,0]
```

OR

```
A=[1 2 -1 2 0;2 1 1 -1 0;3 -1 -2 3 0].
```

Semicolons

Semicolons “;” at the end of a command suppress output to the command window. See the difference with the above commands with and without a semicolon at the end of the line.

Using your own functions

There are three functions that were written for some of the exercises below. You have the ability to write your own, and may be asked to write functions later or in another course. **In order for these “hand-written” functions to run in MATLAB®, they need to be in the folder/directory you are working in**. So you need to pay attention to the “Current Folder” on the MATLAB screen above the editor and command window and/or to the left of the screen. You will need to save these function files in your own folder where you will be saving, running, and possibly publishing, your SCRIPT or LIVE SCRIPT file from. There are some naming conventions to your functions. For example, we cannot name a function `sin.m` since there is already a "sin" function in MATLAB. When MATLAB sees what appears to be a command to run a function, MATLAB always checks for the function file of that name in its own directories first, then the "Current Folder". Thus our `sin.m` function file will never be "found" to run. And if you are in a different folder from where you saved your function file, it will not be found and an error will appear.

RREF

To get a matrix already entered and named A in reduced row echelon form, just type

```
rref(A)
```

Notice that the answer may not give the answer in exact values. You could have MATLAB do this by using the SYMBOLIC MATH TOOLBOX. (Most versions of MATLAB will come with this toolbox.) This is done using the function `sym`. For example, type

```
sym(rref(A))
```

Notice that the resulting output is now in exact value form. REMEMBER THIS; we will be using exact values on multiple exercises.

Another way is to use the `rats` function or change the format of answers to `format rat`. THIS IS ONLY USEFUL IF YOUR ANSWERS AND MATRICES WILL ONLY HAVE INTEGER OR RATIONAL NUMBERS IN THEM. For example, if you switch to `format rat`, notice what happens when you enter `pi`. You can display your matrix fraction, rather than decimal form by entering the following:

```
format rat
rref(A)
```

Or, without changing the format, you could enter:

```
rats(rref(A))
```

To get back to the default format, enter `format` or `format short`. To test, you can always enter `pi` and see if you get `3.1416`.

HELP in MATLAB

If you know a command or function name, you can always get help for it by typing `help command`. For example, you can type `help rref`. Also, if the function files are copied over the current directory you can type `help addrow` to get some help for it. For help with MATLAB defined functions and commands you can also use the Help menu item at the top.

C.1.2 Exercises

These are exercises I've included on the first "introductory" project at various times. I have not included all of them at once. Which exercises have been chosen and how many are chosen depends on the textbook used and the timing of the assignment.

1. Consider the linear system below. Use MATLAB to solve the system by putting it in RREF form but do each step separately (i.e., DON'T use the RREF command). Use the three functions `swap`, `mult`, and/or `addrow` found on the text website. NOTE: In order to be able to use the functions within MATLAB, you MUST copy the three files to your Current Folder within MATLAB.

$$\begin{aligned} x_1 - x_2 - 2x_3 &= -5, \\ 6x_1 - 5x_2 - 7x_3 &= -3, \\ -2x_1 - 6x_3 &= -44. \end{aligned}$$

   You may need to rerun your script file after each step to see what the next step should be. In the end, you should have all steps needed to get it into RREF form. Also, you will need to have exact values (thus the `rats` command may need to be used—at every stage or at the end—you will be the judge).

2. Consider the linear system below. Use MATLAB to solve the system by putting the augmented matrix in RREF form. You may use the `rref` command for this, but make sure that exact answers are given.

$$\begin{aligned} x_1 + 3x_3 + 3x_4 &= -24, \\ x_2 - 4x_3 - 2x_4 &= 29, \\ 3x_1 - 3x_2 + 24x_3 + 15x_4 &= -171, \\ -x_2 + 4x_3 + 7x_4 &= -54. \end{aligned}$$

3. Consider the linear system below. Use MATLAB to solve the system by putting the augmented matrix in RREF form. You may use the `rref` command for this, but make sure that exact answers are given.

$$\begin{aligned} x_1 + x_2 &= -2, \\ x_2 + x_3 &= 2, \\ x_3 + x_4 &= 3, \\ x_1 + x_4 &= -1. \end{aligned}$$

4. Consider the linear system below. Use MATLAB to solve the system by putting the augmented matrix in RREF form. You may use the `rref` command for this, but make sure that exact answers are given.

$$4x_1 - 3x_2 + 3x_3 + 4x_4 = 0,$$

$$\begin{aligned}
-x_1 + x_2 + 2x_3 + 3x_4 &= 5,\\
3x_1 - 2x_2 + 5x_3 + 7x_4 &= 5,\\
-3x_1 + 3x_2 + 6x_3 + 9x_4 &= 15.
\end{aligned}$$

5. Suppose a quadratic polynomial $f(x) = ax^2 + bx + c$ goes through the points $(-1, 3)$, $(0, 2)$, and $(2, 24)$. Using the points, write a system of linear equations to solve for the unknown coefficients of $f(x)$. Use MATLAB to solve this system and thus find the polynomial by putting the augmented matrix in RREF form. You may use the `rref` command for this, but make sure that exact answers are given.

6. Suppose a cubic polynomial $f(x)$ is such that $f(-1) = -6$, $f'(-1) = 2$, $f''(-1) = -4$, and $f'''(-1) = 6$. Using the information about the derivatives, write a linear system to solve for the unknown coefficients of $f(x)$. Use MATLAB to solve this system and thus find the polynomial by putting the augmented matrix in RREF form. You may use the `rref` command for this, but make sure that exact answers are given.

7. Suppose a quartic (degree 4) polynomial $f(x)$ goes through the points $(-2, -33)$, $(-1, 3)$, $(0, 7)$, $(1, 15)$, and $(2, 15)$. Using the points, write a system of linear equations to solve for the unknown coefficients of $f(x)$. Use MATLAB to solve this system and thus find the polynomial by putting the augmented matrix in RREF form. You may use the `rref` command for this, but make sure that exact answers are given.

8. Consider the following matrices:

$$A = \begin{bmatrix} 10 & 10 & 10 & 10 \\ 9 & 8 & 7 & 6 \\ 5 & 4 & 3 & 2 \\ 1 & 1 & 1 & 1 \end{bmatrix}, \qquad B = \begin{bmatrix} 0 & 0 & 1 & 0 \\ 0 & 1 & 0 & 0 \\ 1 & 0 & 0 & 0 \\ 0 & 0 & 0 & 1 \end{bmatrix}.$$

Use MATLAB to calculate AB and BA.

9. In Exercise 8 above you calculated AB and BA. The matrix B is a special matrix in that multiplication with this matrix gives a very specific result. Look closely at your answers for AB and BA.

(a) Multiplying on the **right** by B (so looking at AB) results in an elementary operation performed on the matrix A. Which elementary operation is it?

- **i.** SwapRows. Two rows are swapped.
- **ii.** SwapCols. Two columns are swapped.
- **iii.** MultRow. A row is multiplied by a non-zero constant.
- **iv.** MultCol. A column is multiplied by a non-zero constant.
- **v.** AddRow. A multiple of one row is added to another row.
- **vi.** AddCol. A multiple of one column is added to another column.

(b) Multiplying on the **left** by B (so looking at BA) results in an elementary operation performed on the matrix A. Which elementary operation is it?

i. SwapRows. Two rows are swapped.
ii. SwapCols. Two columns are swapped.
iii. MultRow. A row is multiplied by a non-zero constant.
iv. MultCol. A column is multiplied by a non-zero constant.
v. AddRow. A multiple of one row is added to another row.
vi. AddCol. A multiple of one column is added to another column.

(c) Write a brief description specifically of what the elementary operations are (which rows or columns? multiplied by what?). BE SPECIFIC!

10. Consider the following matrices:

$$C = \begin{bmatrix} 1 & 1 & 1 & 1 \\ 2 & 3 & 4 & 5 \\ 6 & 7 & 8 & 9 \\ 10 & 10 & 10 & 10 \end{bmatrix}, \qquad D = \begin{bmatrix} 1 & 0 & 0 & 0 \\ 2 & 1 & 0 & 0 \\ 0 & 0 & 1 & 0 \\ 0 & 0 & 0 & 1 \end{bmatrix}.$$

Use MATLAB to calculate CD and DC.

11. In Exercise 10 above you calculated CD and DC for the following matrices:

$$C = \begin{bmatrix} 1 & 1 & 1 & 1 \\ 2 & 3 & 4 & 5 \\ 6 & 7 & 8 & 9 \\ 10 & 10 & 10 & 10 \end{bmatrix}, \qquad D = \begin{bmatrix} 1 & 0 & 0 & 0 \\ 2 & 1 & 0 & 0 \\ 0 & 0 & 1 & 0 \\ 0 & 0 & 0 & 1 \end{bmatrix}.$$

The matrix D is a special matrix in that multiplication with this matrix gives a very specific result. Look closely at your answers for CD and DC.

(a) Multiplying on the **right** by D (so looking at CD) results in an elementary operation performed on the matrix C. Which elementary operation is it?

i. SwapRows. Two rows swapped.
ii. SwapCols. Two columns swapped.
iii. MultRow. A row is multiplied by a non-zero constant.
iv. MultCol. A column is multiplied by a non-zero constant.
v. AddRow. A multiple of one row is added to another row.
vi. AddCol. A multiple of one column is added to another column.

(b) Multiplying on the **left** by D (so looking at DC) results in an elementary operation performed on the matrix C. Which elementary operation is it?

i. SwapRows. Two rows are swapped.
ii. SwapCols. Two columns are swapped.
iii. MultRow. A row is multiplied by a non-zero constant.
iv. MultCol. A column is multiplied by a non-zero constant.

v. AddRow. A multiple of one row is added to another row.

vi. AddCol. A multiple of one column is added to another column.

(c) Write a brief description specifically what the elementary operations are (which rows or columns? multiplied by what?). BE SPECIFIC!

For Problems 12 to 18, consider the chemical reaction given where a, b, c, and c are unknown positive integers. The number of atoms of each element must be the same before and after the reaction for the reaction to be balanced. For example, because the number of oxygen atoms must remain the same, for Exercise 12 below we have the equation

$$2a + b = 2c + 3d.$$

For each element in the reaction, we get an equation that together form a linear system representing the chemical reaction. Many times there are not enough equations to get an exact solution; thus there may be infinitely many solutions overall for the variables. Typically, one gives the solution that gives the smallest possible positive integers. Balance the reactions using the smallest possible valid positive integers for your answers.

(a) Write the system of equations to balance the chemical reaction.

(b) Write the augmented matrix used to solve this system of equations.

(c) Use the `rref` command on the augmented matrix within MATLAB to find solution(s) to the system. You may want to use `sym(rref(A))` so the answers appear in exact form rather than numerical approximations. Write the general solution to the system. Then, find the solution for balancing the equation; that is, state the solution that uses the smallest positive integers possible to balance the equation.

(d) Write the system of equations in the form of $Ax = 0$, stating what A and x equal for this system. Use the "divide into" operator within MATLAB `A\b` to solve the system $Ax = b$ for x. Does this give you the desired answer?

(e) Use the commands `null(A)` and `null(sym(A))`.

12. The chemical reaction of nitrogen dioxide and water into nitrous and nitric acids:

$$aNO_2 + bH_2O \rightarrow cHNO_2 + dHNO_3.$$

13. The chemical reaction of phosphorus pentoxide and calcium fluoride into phosphorus pentafluoride and tricalcium phosphate:

$$aP_4O_{10} + bCaF_2 \rightarrow cPF_5 + dCa_3(PO_4)_2.$$

14. The chemical reaction of aluminum and water into aluminum hydroxide and hydrogen:

$$aAl + bH_2O \rightarrow cAl(OH)_3 + dH_2.$$

15. The chemical reaction of ammonia and oxygen into nitric oxide and water:

$$aNH_3 + bO_2 \rightarrow cNO + dH_2O.$$

16. The chemical reaction hydrazine and dinitrogen tetroxide into nitrogen and water:

$$aN_2H_4 + bN_2O_4 \rightarrow cN_2 + dH_2O.$$

17. The chemical reaction of methane and oxygen into carbon dioxide and water:

$$aCH_4 + bO_2 \rightarrow cCO_2 + dH_2O.$$

18. The chemical reaction of ethane, carbon dioxide, and water into ethanol:

$$aC_2H_6 + bCO_2 + cH_2O \rightarrow dC_2H_5OH.$$

C.2. The Hill cipher

The Hill cipher cannot be decoded using the same techniques as the basic substitution or transposition ciphers. Those techniques include counting the number of times letters are used to narrow down the code for common letters in the English language, etc. The Hill cipher is such that the same letter may appear differently in each occurrence in the message. Thus messages encoded using the Hill cipher can be very difficult to decode by hand without knowing a "key". We will not discuss the pros and cons of the Hill cipher.

The Hill cipher takes strings of letters in a message, translates them into numbers (A = 0, B = 1, ..., Z = 25) and puts these numbers in a column of size m, creating a matrix of size $m \times n$ ($m =$ the number of columns) and multiplies them by an $m \times m$ matrix called a key. If the message to encode does not fill up a column, you can use "dummy letters" to complete the column. Once multiplied by the key, you end up with a new $m \times n$ matrix that you then change back into your encoded message. This is done by reducing the numbers MODULO 26 and translating the numbers back into letters.

Example C.2.1. Use the key $K = \begin{bmatrix} 2 & 3 \\ 1 & 4 \end{bmatrix}$ to encode "Linear Algebra" using the letter "Z" for any "dummy letters" that are needed.

Notice we have a 2×2 key so we will be splitting the letters up into pairs, so the columns will be of size $m = 2$. We have an odd number of letters so we will use Z $= 25$ for the "dummy letter" at the end and we will have seven pairs (so $n = 7$).

$$\text{LINEAR ALGEBRA} = 11\ 8\ 13\ 4\ 0\ 17\ 0\ 11\ 6\ 4\ 1\ 17\ 0$$

Thus the 2×7 matrix becomes

$$M = \begin{bmatrix} 11 & 13 & 0 & 0 & 6 & 1 & 0 \\ 8 & 4 & 17 & 11 & 4 & 17 & 25 \end{bmatrix}.$$

Multiply the key with this matrix:

$$\begin{aligned} KM &= \begin{bmatrix} 2 & 3 \\ 1 & 4 \end{bmatrix} \begin{bmatrix} 11 & 13 & 0 & 0 & 6 & 1 & 0 \\ 8 & 4 & 17 & 11 & 4 & 17 & 25 \end{bmatrix} = \begin{bmatrix} 46 & 38 & 51 & 33 & 24 & 53 & 75 \\ 43 & 29 & 68 & 44 & 22 & 69 & 100 \end{bmatrix} \\ &= E. \end{aligned}$$

Now we need to translate this new 2×7 matrix back into letters. This is where MODULAR arithmetic comes in. We have used modular arithmetic without even realizing it; translating the 24-hour clock into the 12-hour clock is arithmetic modulo 12. We may also do this when giving change to a cashier to get only quarters back. Here is a more formal mathematical definition: Let a, b, and n be integers. We say a is equivalent to b modulo n, denoted $a \equiv b \pmod{n}$, if $a - b$ is an integer multiple of n. When dealing with positive integers, the easiest way to think about it is, "What is the remainder when we divide a by n? That is b." So for instance, when I divide 16 by 12, the remainder is 4 so $16 \equiv 4 \pmod{12}$. When I divide 88 by 25, I get a remainder of 13 so $88 \equiv 13 \pmod{25}$. For modulo 26, $26 \equiv 0 \pmod{26}$, $27 \equiv 1 \pmod{26}$, $28 \equiv 2 \pmod{26}$ and so on.

THERE IS A COMMAND IN MATLAB FOR MODULAR ARITHMETIC; THIS WILL BE USEFUL!

Thus for the matrix E above we have

$$E \equiv \begin{bmatrix} 20 & 12 & 25 & 7 & 24 & 1 & 23 \\ 17 & 3 & 16 & 18 & 22 & 17 & 22 \end{bmatrix} \pmod{26},$$

which then transforms into the letters `URMDZQHSYWBRXW`. Thus "LINEAR ALGEBRA" gets encoded into "URMDZQHSYWBRXW". Notice how the letter "A" is encoded differently each time in our message.

DECRYPTION:

Obviously we would want to be able to decrypt a Hill cipher. This would involve solving $KM = E$ where M is unknown but E is the encoded message we are given and K was the original encryption key matrix. Thus, if K^{-1} exists and we knew what it equaled, we would get $M = K^{-1}E$. So we would want K to be an invertible matrix, or a nonsingular one. We also want K^{-1} to have entries that are integers, modulo 26.

There are other issues because of the modular arithmetic that we will not get into right now. So how to do this?

For the above key $K = \begin{bmatrix} 2 & 3 \\ 1 & 4 \end{bmatrix}$, we get

$$K^{-1} = \begin{bmatrix} \frac{4}{5} & -\frac{3}{5} \\ -\frac{1}{5} & \frac{2}{5} \end{bmatrix}.$$

We do not want fractions so what are these fractions equivalent to modulo 26? The easiest way to show you is by example. Remember $26 \equiv 0 \pmod{26}$ so $\frac{26}{5} \equiv \frac{0}{5} \pmod{26}$. Thus

$$\frac{4}{5} + \frac{26}{5} = \frac{30}{5} = 6$$

and

$$\frac{-3 + 26 + 26 + 26}{5} = \frac{75}{5} = 15.$$

Thus $\frac{4}{5} \equiv 6 \pmod{26}$ and $-\frac{3}{5} \equiv 15 \pmod{26}$. So you get a decryption matrix $D \equiv K^{-1} \pmod{26}$ and for our example,

$$D = \begin{bmatrix} 6 & 15 \\ 5 & 16 \end{bmatrix}.$$

Notice (check it!), if we would be given the encoded message E above, we get

$$DE \equiv M \quad (\text{mod } 26).$$

TIP: If converting fractions modulo 26 by hand, write out a bunch of multiples of 26. By writing out a bunch of multiples, I saw that $3 \cdot 26 = 78$ so if I subtracted 3, I get a nice multiple of 5, etc.

Notice that we are not encoding numbers, spaces or punctuation—this can be done as well by adding to our "alphabet". For example, we could have a space = 26, "." = 27, and "?" = 28 and we would be working modulo 29.

C.2.1 Useful commands

In computers, **ASCII codes** represent text. For the upper-case English alphabet, the codes are from 65–90. Thus the ASCII code for A is 65 and the ASCII code for Z is 90. You can look up ASCII codes for other text and symbols but these are the only ones we are using for this project. We are going to use the following commands to convert between letters and numbers. There are also some useful commands to convert vectors to matrices, etc. that will be useful to us that is discussed below.

NOTE: Not all of the details of these commands are discussed here. Only an explanation of what we need to know about the commands for our use for this project is provided. Feel free to explore these commands further on your own!

DOUBLE

The command we need to convert our strings to ASCII codes is `double`. Here is an example:

```
>> x='ABCD'
x =
    'ABCD'
>> y=double(x)
y =
    65    66    67    68
```

CHAR

The command to convert numbers that are ASCII codes to text is `char`. Here is an example.

```
>> x=0:25; % creates a vector of the numbers from 0 to 25.  Output is suppressed.
>> y=x+65 % adds 65 to every element in the vector so the is now from 65 to 90.
y =
  Columns 1 through 18
    65    66    67    68    69    70    71    72    73    74    75    76    77    78
            79    80    81    82
  Columns 19 through 26
    83    84    85    86    87    88    89    90
>> alphabet=char(y) % alphabet is now a string that should all uppercase.
alphabet =
    'ABCDEFGHIJKLMNOPQRSTUVWXYZ'
```

RESHAPE

We want to "reshape" a vector X to a matrix of size $m \times n$. Then we use the command `reshape(X,m,n)`:

```
>> ex1 = reshape(alphabet, 2, 13)
ex1 =
  2{\texttimes}13 char array
    'ACEGIKMOQSUWY'
    'BDFHJLNPRTVXZ'
>> A=[1 2 3 4]
A =
```

```
     1     2     3     4
>> reshape(A,2,2)
ans =
     1     3
     2     4
```

MOD

We want to see what numbers are equivalent modulo 26. The command `mod(x,26)` will do this. Let us look at some examples. For example, we know from clocks that 14 is equivalent to 2 modulo 12:

```
>> mod(14,12)
ans =
     2
```

We can also use the `mod` command if we want the last four digits of a number. Use `mod(x,10000)`:

```
>> mod(987654321,10000)
ans =
        4321
```

For us, we want matrices modulo 26. Luckily, the command works on vectors and matrices.

```
>> x = [12, 30; 46 53]
x =
    12    30
    46    53
>> y=mod(x,26)
y =
    12     4
    20     1
```

Since 30, 46, and 53 are equivalent to 4, 20, and 1, respectively, modulo 26 we see that `y` now has the values 12, 4, 20, and 1.

Using MATLAB for the Hill cipher example

What follows is an explanation of how to use MATLAB to do the work for us on the Hill Cipher handout.

Our key is the matrix K:

```
K = [2 3;1 4];
```

The numbers for our message are LINEARALGEBRA = 11 8 13 4 0 17 0 11 6 4 1 17 0. We could have figured this out on paper (which is error-prone) or we can use MATLAB. For the project, we want to use MATLAB as much as possible.

```
>> msg='LINEARALGEBRA'
msg =
    'LINEARALGEBRA'
>> msgNumber=double(msg)
msgNumber =
    76    73    78    69    65    82    65    76    71    69    66    82    65
```

Notice that `msgNumber` is now a vector of the ASCII codes for our message. For the Hill Cipher we want the numbers to be from 0 to 25 instead of 65 to 90. Thus we will subtract 65 from our `msgNumber`.

```
>> msgNumber=msgNumber-65
msgNumber =
    11     8    13     4     0    17     0    11     6     4     1    17     0
```

Now we need to get these numbers into the correct format. Notice that there are 13 letters in this message, we want 2 rows so we would have seven columns and thus we need one "dummy letter" of $Z = 25$ at the end. I could add this at the end of my `msgNumber` vector (if you know how to do this) or now that I've figured it out, I just redo the above commands. In your script file you would just correct the original commands you had.

```
>> msg='LINEARALGEBRAZ'
msg =
    'LINEARALGEBRAZ'
>> msgNumber = double(msg)-65
msgNumber =
    11     8    13     4     0    17     0    11     6     4     1    17     0    25
```

Notice that these numbers are the same as on the handout. But from these numbers we need to create the 2 × 7 matrix M. Thus we use the `reshape` command:

```
>> msgNumberMtx=reshape(msgNumber,2,7)
msgNumberMtx =
    11    13     0     0     6     1     0
     8     4    17    11     4    17    25
```

Now we do the encoding. When we do the matrix multiplication we see that we do not get numbers between 0 and 25 so we use the `mod` command (discussed above). Then we transfer the matrix back to a vector and then letters. Notice that the encoded message is the same as on the handout. I did not suppress any output so you could see what each command does. You may want to suppress the output from some or all of the intermediate steps so that you only see the end result.

```
>> encodedMsg=K*msgNumberMtx
encodedMsg =
    46    38    51    33    24    53    75
    43    29    68    44    22    69   100
>> encodedMsg=mod(encodedMsg,26)
encodedMsg =
    20    12    25     7    24     1    23
    17     3    16    18    22    17    22
>> encodedMsgNum=reshape(encodedMsg,1,14)
encodedMsgNum =
    20    17    12     3    25    16     7    18    24    22     1    17    23    22
>> encodedMsgNum=encodedMsgNum+65
encodedMsgNum =
    85    82    77    68    90    81    72    83    89    87    66    82    88    87
>> codedMsg=char(encodedMsgNum)
codedMsg =
    'URMDZQHSYWBRXW'
```

Finding the inverse of the key

There are several ways one can find an inverse of a matrix in MATLAB. We could do it in MATLAB the way we do it my hand or by using the command `inv`. We are safe to use the `rats` command since we know our original matrix has only integer entries. It would NOT be safe if our original matrix had irrational entries.

```
>> D=inv(K)
D =
    0.8000   -0.6000
   -0.2000    0.4000
>> rats(D)   % this is for display purposes only
ans =
  2{\texttimes}28 char array
    '      4/5           -3/5      '
    '     -1/5            2/5      '
```

Now we need to figure out what these fractions are modulo 26.

```
>> a=26/5;       % Didn't want to type 26/5 over and over
>> D(1,1)+a      % Ran once, saw it was an integer between 0 and 26 so done
ans =
     6
>> D(1,1)=ans  % Reassigning the 1,1 entry of D to be the above answer
D =
    6.0000   -0.6000
```

```
   -0.2000    0.4000
>> D(1,2) + a + a + a % Re-ran this, adding "a" each time until get desired integer
ans =
    15
>> D(1,2)= ans  % Reassigning the 1,2 entry of D to be the previous answer
D =
    6.0000   15.0000
   -0.2000    0.4000
>> D(2,1)+a
ans =
     5
>> D(2,1)=ans
D =
    6.0000   15.0000
    5.0000    0.4000
>> D(2,2)+a+a+a
ans =
    16
>> D(2,2)=ans
D =
     6    15
     5    16
```

Now we have the inverse matrix, modulo 26 as in the handout. Let us check against our encoded message. Output of the intermediate steps is suppressed.

```
>> codedMsg
codedMsg =
    'URMDZQHSYWBRXW'
>> numCode2 = double(codedMsg);
>> numCode2=numCode2-65;    % convert to numbers from 0 to 25
>> E2=reshape(numCode2,2,7) % reshape to 2x7 matrix
E2 =
    20    12    25     7    24     1    23
    17     3    16    18    22    17    22
```

Notice E2 is the same as E in the above example. Now decode and then convert to letters:

```
>> chkMsg=D*E2;
>> chkMsg=mod(chkMsg,26);
>> chkMsg=reshape(chkMsg,1,14);
>> chkMsg=chkMsg+65;
>> char(chkMsg)
ans =
    'LINEARALGEBRAZ'
```

We may have been told that the dummy letter(s) that would be used would be Z, or we see that the entire decoded message does not make sense as given and recognize that the original message was "LINEAR ALGEBRA".

C.2.2 Exercises

Complete these problems within MATLAB, showing all work, even calculations done, within MATLAB.

1. Consider the Hill cipher with the key

$$\begin{bmatrix} 2 & 5 \\ 1 & 4 \end{bmatrix}.$$

   (a) Encode the following text, adding Z if necessary, stripping spaces and punctuation:
   Step on no pets.
   (b) Calculate the "inverse matrix" (decode matrix) with entries that are integers from 0 to 25 (integers modulo 26).
   (c) Using the decode matrix above, decode the following message, stripping off any extraneous "dummy letters" that may have been put at the end. In other words, strip off any extraneous "dummy letters" so that the decoded message is an actual word or phrase in English.
   SIEBZY

2. Consider the Hill cipher key in Exercise 1 above and its decode matrix.
   (a) Use the key to encode "Name no one man," adding Q if necessary, stripping spaces and punctuation.
   (b) Use the decode matrix to decode "IRGDTF" stripping off any extraneous "dummy letters" that may have been put at the end.

3. Consider the Hill cipher with the key

$$\begin{bmatrix} 1 & 5 \\ 2 & 7 \end{bmatrix}.$$

   (a) Encode the following text, adding X if necessary, stripping spaces and punctuation:
   Madam, I'm Adam.
   (b) Calculate the "inverse matrix" (decode matrix) with entries that are integers from 0 to 25 (integers modulo 26).
   (c) Using the decode matrix above, decode the following message, stripping off any extraneous "dummy letters" that may have been put at the end. In other words, strip off any extraneous "dummy letters" so that the decoded message is an actual word or phrase in English.
   FYPSNE

4. Consider the Hill cipher key in Exercise 3 above and its decode matrix.
  (a) Use the key to encode "Was it a rat I saw?," adding Q if necessary, stripping spaces and punctuation.
  (b) Use the decode matrix to decode "KUYWFN" stripping off any extraneous "dummy letters" that may have been put at the end.

C.3. Least-squares solutions

C.3.1 Brief overview

This should supplement a section in a linear algebra text on least-squares solutions as many details are not provided here.

When the system $A\mathbf{x} = \mathbf{b}$ has no solution, we may want to find an approximation to a solution. This is usually when we have too many equations, and the system is **overdetermined**. Since $A\mathbf{x} - \mathbf{b} \neq \mathbf{0}$ for any $\mathbf{x}$, we want to look at approximate solutions $\widehat{\mathbf{x}}$ and $A\widehat{\mathbf{x}} - \mathbf{b} = \mathbf{e}$, the "error" vector. If we make the length of $\mathbf{e}$ as small as possible, we've minimized the error and have a "best approximation" to a solution to $A\mathbf{x} = \mathbf{b}$. This is what a **least-squares solution** is; the vector $\widehat{\mathbf{x}}$ that makes

$$\|\mathbf{e}\| = \|A\widehat{\mathbf{x}} - \mathbf{b}\|$$

as small as possible.

Without getting into the details, we find the least-squares solution $\widehat{\mathbf{x}}$ to $A\mathbf{x} = \mathbf{b}$ by solving the system $A^T A\mathbf{x} = A^T\mathbf{b}$.

Example C.3.1. Consider the system

$$\begin{aligned} 5x + 6y &= 3, \\ 6x + 7y &= 1, \\ x + y &= -5. \end{aligned}$$

```
>> A=[5 6;6 7;1 1]; b=[3;1;-5];
>> M=[A b]
M =
     5     6     3
     6     7     1
     1     1    -5
>> rref(M)
ans =
     1     0     0
     0     1     0
     0     0     1
```

We see above that the system has no solutions. But the least-squares solution is the solution to $A^T A\mathbf{x} = A^T\mathbf{b}$:

```
>> B=A'*A; c=A'*b;
>> N=[B c]
N =
    62    73    16
    73    86    20
>> rref(N)
ans =
     1     0   -28
     0     1    24
```

So the least-squares solution to $A\mathbf{x} = \mathbf{b}$ is

$$\hat{\mathbf{x}} = \begin{bmatrix} -28 \\ 24 \end{bmatrix}.$$

C.3.2 Curve fitting

A common use for least-squares solutions is curve fitting.

Example C.3.2. Suppose we want a parabola (quadratic function) that best fits these data points:

$$(-1,\ 1) \quad (3,\ 0) \quad (0,\ 1) \quad (-2,\ -2) \quad (2,\ 3)$$

By looking at $f(x) = ax^2 + bx + c$ we get the following equations:

$$\begin{aligned}
f(-1) = 1 &\implies & a - b &\ +c = 1,\\
f(3) = 0 &\implies & 9a + 3b &\ +c = 0,\\
f(0) = 1 &\implies & &\ c = 1,\\
f(-2) = -2 &\implies & 4a - 2b &\ +c = -2,\\
f(2) = 3 &\implies & 4a + 2b &\ +c = 3.
\end{aligned}$$

These equations form a linear system $A\mathbf{x} = \mathbf{b}$:

```
>> A=[1 -1 1;9 3 1;0 0 1;4 -2 1;4 2 1];
>> b=[1;0;1;-2;3];
>> M=[A b]
M =
     1    -1     1     1
     9     3     1     0
     0     0     1     1
     4    -2     1    -2
```

```
     4     2     1     3
>> rref(M)
ans =
     1     0     0     0
     0     1     0     0
     0     0     1     0
     0     0     0     1
     0     0     0     0
```

We see above that there is no solution. We find a least-squares solution:

```
>> rref([A'*A A'*b])
ans =
    1.0000         0         0   -0.5000
         0    1.0000         0    1.0000
         0         0    1.0000    2.0000
>> rats(rref([A'*A A'*b]))
ans =
  3{\texttimes}56 char array
    '       1          0          0         -1/2      '
    '       0          1          0          1        '
    '       0          0          1          2        '
```

So the least-squares solution is $\widehat{x} = \begin{bmatrix} -\frac{1}{2} \\ 1 \\ 2 \end{bmatrix}$ and thus the quadratic function that best fits the data is $f(x) = -\dfrac{1}{2}x^2 + x + 2$ (see Fig. C.1).

C.3.3 Exercises

All of the work to answer the following exercises must be done in MATLAB and appear in your file. Even basic calculations that you would normally need scratch paper or your calculator for should be done in MATLAB.

1. Consider the following system:

$$\begin{bmatrix} 2 & -2 \\ -2 & 2 \\ 5 & 5 \end{bmatrix} \mathbf{x} = \begin{bmatrix} 15 \\ -9 \\ 10 \end{bmatrix}.$$

(a) Use MATLAB to try and solve the system and notice why we may need least-squares.

(b) Use MATLAB to find the least-squares solution $\widehat{x}$ of the system.

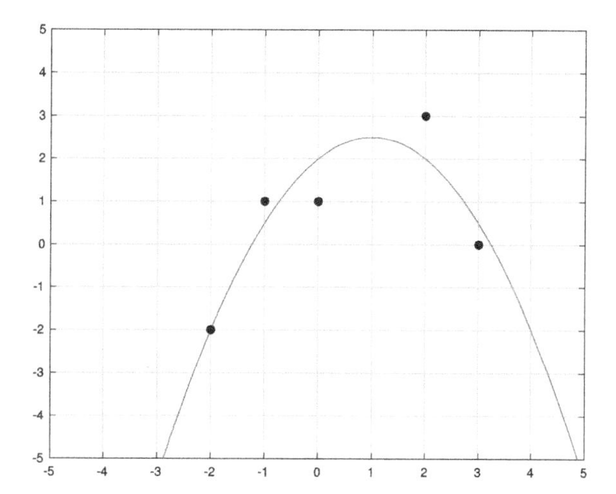

Figure C.1 Data with curve fitting.

2. Consider the following system:

$$\begin{bmatrix} 1 & 1 & -1 \\ 1 & 1 & 1 \\ 1 & -1 & 1 \\ 1 & -1 & -1 \end{bmatrix} \mathbf{x} = \begin{bmatrix} 7 \\ 7 \\ -7 \\ 5 \end{bmatrix}.$$

   (a) Use MATLAB to try and solve the system and notice why we may need least-squares.

   (b) Use MATLAB to find the least-squares solution $\widehat{x}$ of the system.

3. Consider the linear function $f(x) = mx + b$ with the data points $(-2, 8)$, $(0, 4)$, and $(2, 12)$.

   (a) Come up with a linear system to solve for m and b.

   (b) Use MATLAB to try and solve the system and notice why we may need least-squares.

   (c) Use MATLAB to find the least-squares solution $\widehat{x}$ of the system.

   (d) BONUS: plot the data points and the line $y = mx + b$ on the same figure in MATLAB.

4. Consider the quadratic polynomial $f(x) = ax^2 + bx + c$ with the data points $(0, -1)$, $(1, 0)$, $(2, -7)$, and $(3, -2)$.

   (a) Come up with a linear system to solve for a, b, and c.

   (b) Use MATLAB to try and solve the system and notice why we may need least-squares.

   (c) Use MATLAB to find the least-squares solution $\widehat{x}$ of the system.

(d) BONUS: plot the data points and the line $y = f(x)$ on the same figure in MATLAB.

C.4. Markov matrices

C.4.1 Brief overview

One use for eigenvalues and eigenvectors is with **Markov matrices**. There are many texts on this subject; for example, Section 2.8 in Ref. [27] discusses Markov matrices in more detail with several examples. There is a lot to study about Markov matrices and Markov Chains; we will only introduce the idea.

In our application we have a **Markov matrix** without going into detail of the definition of a Markov matrix. A Markov matrix is also called a stochastic matrix, a transition matrix, a probability matrix, or a substitution matrix. The type of Markov matrix we will be looking at is a matrix in which:

- All entries of the matrix are nonnegative.
- The entries of each column add up to 1.

In our examples, the entries of the Markov matrix are probabilities that something will *transition* from one state to another. Thus the entry p_{ij} is the probability that of transitioning from state i to state j in one "generation."

Consider the Markov matrix

$$P = \begin{bmatrix} p_{11} & p_{12} \\ p_{21} & p_{22} \end{bmatrix}$$

where one "generation" is one day. Then the value of p_{11} is the probability someone in group A will stay in group A the next day, p_{12} is the probability someone in group A will go to group B, p_{21} is the probability that someone in group B will go to group A the next day, and finally p_{22} is the probability that someone in group B will stay in group B the next day.

In this project we will look at two examples and investigate the relationship between the Markov matrices, eigenvalues, and eigenvectors. We will also have to remember how to calculate limits from calculus. It may be useful to remind yourself of the limit of a **geometric sequence**.

There will be work shown in your MATLAB file and maybe some work on paper to do these problems. There are examples of how to work with the characteristic function and/or find eigenvalues on the text website.

C.4.2 Exercises

All of the work to answer the exercises below must be done in MATLAB and appear in your file. Even basic calculations that you would normally need scratch paper or your calculator for should be done in MATLAB. **USING CALCULATORS, MATHEMATICA, MAPLE, WOLFRAM ALPHA, ETC. IS NOT ALLOWED.** Any written work should be handed in.

1. At the end of every month a manager rates the performance of each member of their staff as poor, average, or excellent. If a worker was rated poor at the end of one month, then the probability that the next month the worker will be rated poor is 0.15, average is 0.7 and excellent is 0.15. If a worker was rated average at the end of one month then probability that the next month the worker will be rated poor and average is 0.2 and 0.5, respectively. If a worker was rated excellent at the end of one month then the probability that the next month the worker will be rated poor is 0.05 and average is 0.75.

Complete the construction of the Markov matrix A below for the above information. The first column of the matrix are the probabilities of being rated poor at the end of one month and then rated the next month as poor, average and excellent, respectively. Thus the first column of the matrix looks like

$$\begin{bmatrix} 0.15 \\ 0.7 \\ 0.15 \end{bmatrix}.$$

The second column of A has the probabilities of being rated average at the end of one month and then being rated the next month as poor, average and excellent, respectively and so on. Recall that probabilities should add up to 1, that is that at the end of any given month, the probabilities of being rated poor, average, or excellent should add up to 1.

Find A. If you need calculations done to figure out the entries, those should be done in MATLAB. At the minimum, the only thing needed to be done in the MATLAB is to enter the matrix, named A.

2. Use the Markov matrix A from the previous exercise.

(a) Suppose that initially, 10% of the workers were rated poor and 75% of the workers were rated as average. Complete the entries for the initial state vector $\mathbf{x}_0$, stating the values as percentages (thus 16 rather than 0.16 for 16%).

(b) Use MATLAB to find the **percentage** of workers will be rated poor after 5 months. Round your answer to two decimal places.

3. Use the Markov matrix from Exercise 1.

(a) Find the characteristic polynomial $p(x)$ of that matrix:

(b) Given that a Markov matrix always has an eigenvalue of $\lambda = 1$, find the other 2 eigenvalues with $\lambda_2 < \lambda_3$. **USING CALCULATORS, MATHEMATICA, MAPLE, WOLFRAM ALPHA, ETC. IS NOT ALLOWED.** State the exact values of the eigenvectors.

4. Suppose you had the following Markov matrix instead of the one given in the previous exercises.

$$\begin{bmatrix} 0.2 & 0.3 & 0.1 \\ 0.7 & 0.4 & 0.7 \\ 0.1 & 0.3 & 0.2 \end{bmatrix}.$$

The eigenvalues for this Markov matrix are $\lambda_1 = 1$, $\lambda_2 = -\frac{3}{10}$, and $\lambda_3 = \frac{1}{10}$, with corresponding eigenvectors

$$\mathbf{u}_1 = \begin{bmatrix} 1 \\ \frac{7}{3} \\ 1 \end{bmatrix}, \quad \mathbf{u}_2 = \begin{bmatrix} 1 \\ -2 \\ 1 \end{bmatrix}, \quad \mathbf{u}_3 = \begin{bmatrix} -1 \\ 0 \\ 1 \end{bmatrix},$$

respectively. Notice that the matrix is NOT defective. Recall the theorem in linear algebra that says these eigenvectors are linearly independent. Thus these vectors form a basis for $\mathbb{R}^3$ so if our initial state vector was

$$\mathbf{x}_0 = \begin{bmatrix} 15 \\ 75 \\ 10 \end{bmatrix},$$

we can write this initial state vector as a linear combination of the three eigenvectors. In other words, we can write

$$\mathbf{x}_0 = a_1\mathbf{u}_1 + a_2\mathbf{u}_2 + a_3\mathbf{u}_3.$$

(a) Use MATLAB to solve for the coefficients a_1, a_2, and a_3.

(b) Write the exact values for these coefficients.

(c) Since we have a difference equation $\mathbf{x}_k = A^k\mathbf{x}_0$, we can write

$$\mathbf{x}_k = A^k(a_1\mathbf{u}_1 + a_2\mathbf{u}_2 + a_3\mathbf{u}_3).$$

Simplify the right hand side of this equation to be in terms of the coefficients above, eigenvalues, and eigenvectors. Use the exact values for the coefficients and eigenvalues but keep $\mathbf{u}_1, \mathbf{u}_2, \mathbf{u}_3$ in the equation.

(d) Write the resulting equation.

(e) The above equation can make calculations easier, and thus faster. Based on this equation, use MATLAB to find the percentage that was rated as **average** after 30 months, both the exact value and numerical approximation rounded to four decimal places.

(f) Based on the equation from the second part (with the exact values), figure out what the $\lim_{k \to \infty} \mathbf{x}_k$ equals using exact values. What is your answer rounded to four decimal places? (use MATLAB)

(g) Based on the meaning behind the Markov matrix and vectors $\mathbf{x}_0$ and $\mathbf{x}_k$, explain what this limit means for this particular application.

APPENDIX D

Multivariable Calculus Projects

D.1. Lines and planes

Copy the template script file to your own folder. Change the equations, etc. to be the equations for your problems. You may need to adjust the DOMAINS to get better pictures! For # 2 and # 3: you are creating a 3D picture of these lines, planes, etc. Then you may use the function `AnimateView` found on the text website to create an animated GIF file of "spinning" the figure around.

1. Consider the line $L(t) = \langle 2+t, 7+5t \rangle$.
   - **(a)** What is the point where the line L intersects the x-axis? What is the value of t for that point?
   - **(b)** What is the point where the line L intersects the y-axis? What is the value of t for that point?
   - **(c)** At what point(s) does the line L intersect the parabola $y = x^2$? What is the value of t for the point(s)?

   Plot the line L, the points of intersection and the parabola $y = x^2$ on the same figure in MATLAB®. Create an appropriate legend and title.
2. Consider the two lines $L_1 : \langle -2t,\ 1+2t,\ 3t \rangle$ and $L_2 : \langle -5+s,\ 2+3s,\ 2+4s \rangle$. Find the point of intersection of the two lines. Plot the two lines and the point of intersection on the same figure in MATLAB.

   BONUS: create an animated GIF that by using the `AnimateView` function found on the text website.
3. Consider the plane $-0.5x - 2.5y + z = 1$.
   - **(a)** Find the parametric equations for the line through the point $P = (5, 3.75, 2.75)$ that is perpendicular to the plane where the parameter $t = 0$ should correspond to the point P. The direction vector of the line should be the same as the standard normal vector of the plane.
   - **(b)** Find the point Q where the line intersects the yz-plane.

   Plot the plane and line in MATLAB on the same figure. Also mark points P and Q on the figure.

   BONUS: create an animated GIF that by using the `AnimateView` function found on the text website.

D.2. Vector functions

D.2.1 2D example plots

In 2D, vector functions are not different from the parametric equations you saw in single variable calculus. To plot them, you must first establish your domain for t. The easiest way to do this is with the command `linspace`.

```
t = linspace(0,2*pi);
```

IMPORTANT: notice the semi-colon at the end of these lines; if there is none, all of those values for t (and x, y, etc. below) will appear in your command window. Now we enter the equations for x and y, using **component-wise** calculations.

```
x = 3*cos(t);
y = 2*sin(t);
```

Now we can plot the x and y values. For 2D plots, it is simple:

```
plot(x,y)
title('2D Vector Function Example 1')
```

The above equations did not need **component-wise** calculations. This example does: $\mathbf{r}(t) = \left\langle \dfrac{\sin(2t)}{4+t^2}, \dfrac{\cos(2t)}{4+t^2} \right\rangle$

```
t = linspace(0,10);
x = sin(2*t)./(4 + t.^2);
y = cos(2*t)./(4 + t.^2);
plot(x,y)
title('2D Vector Function Example 2')
```

D.2.2 3D example plot

This example involves **component-wise** calculations, and the domain needed to be adjusted to show the entire trefoil knot. Notice for a 3D space curve, the command is `plot3` instead of `plot`. It is always good to label the axes for perspective.

```
t=linspace(0,4*pi);
x=(2+cos(1.5*t)).*cos(t);
y=(2+cos(1.5*t)).*sin(t);
z=sin(1.5*t);
plot3(x,y,z)
xlabel('x'), ylabel('y'), zlabel('z')
title('3D Vector Function/Space Curve Example')
```

D.2.3 Bad domain example

The following is an example in which the domain does not have enough points and it creates a jagged curve (try it yourself!) that should be smooth.

```
t=linspace(0,4*pi);
x=cos(6*t);
y=sin(6*t);
z=t;
plot3(x,y,z)
xlabel('x'), ylabel('y'), zlabel('z')
title('Bad Domain Example')
```

The problem above was because by default the `linspace(a,b)` command creates one hundred values between a and b for MATLAB use to connect to plot the graph. By having `linspace(a,b,n)`, you are specifying n values between a and b. So below, we have t equal a vector of 500 values between 0 and 4π with which to create the x, y, and z values to build points to connect into a graph.

```
t=linspace(0,4*pi,500);
x=cos(6*t);
y=sin(6*t);
z=t;
plot3(x,y,z)
xlabel('x'), ylabel('y'), zlabel('z')
title('Better Domain Example')
```

D.2.4 Adjusting the view

You can adjust the view. Using this command may involve experimentation with the values and rerunning the script.

```
t=linspace(0,4*pi);
x=(2+cos(1.5*t)).*cos(t);
y=(2+cos(1.5*t)).*sin(t);
z=sin(1.5*t);
plot3(x,y,z)
view(0,90)
xlabel('x'), ylabel('y'), zlabel('z')
title('Adjusting View on 3D Graph')
```

D.2.5 Sphere command

The `sphere` command will set up variables x, y, and z to form a unit sphere using the `surf` or `mesh` commands. Without going into details, here are examples.

```
[x,y,z]=sphere(100);
mesh(x,y,z)
xlabel('x'), ylabel('y'), zlabel('z')
title('Sphere Command: Unit Sphere')
axis equal
axis([-5 5 -5 5 -5 5])
```

How to get one of a different radius ($r = 3$) than 1:

```
[x,y,z]=sphere(100);
mesh(3*x,3*y,3*z)
xlabel('x'), ylabel('y'), zlabel('z')
title('Sphere with different radius')
axis equal
axis([-5 5 -5 5 -5 5])
```

How to get a sphere with a different center $C(2, -1, 3)$ from the origin:

```
[x,y,z]=sphere(100);
mesh(x+2,y-1,z+3)
xlabel('x'), ylabel('y'), zlabel('z')
title('Sphere with center not at the origin')
axis equal
axis([-5 5 -5 5 -5 5])
```

QUESTION: How could you get ANY sphere; a different center from the origin and radius other than 1?

Here they are side-by-side for comparison:

```
subplot(131)
mesh(x,y,z)
xlabel('x'), ylabel('y'), zlabel('z')
title('Sphere Command: Unit Sphere')
axis equal
axis([-5 5 -5 5 -5 5])
subplot(132)
mesh(3*x,3*y,3*z)
xlabel('x'), ylabel('y'), zlabel('z')
title('Sphere with different radius')
axis equal
axis([-5 5 -5 5 -5 5])
subplot(133)
mesh(x+2,y-1,z+3)
xlabel('x'), ylabel('y'), zlabel('z')
title('Sphere with center not at the origin')
axis equal
axis([-5 5 -5 5 -5 5])
```

D.2.6 Multiple plots on one figure

You must use the `hold on` and `hold off` commands around the additional `plot/plot3/mesh`, etc. commands for all to show up within the same figure.

```
t = linspace(-pi,3*pi);
x = 2*cos(t);
y = sin(t);
z = t;
plot3(x,y,z, 'k')
xlabel('x'), ylabel('y'), zlabel('z')
title('Multiple Graphs in One Example')
hold on           % use this to add more plots to current figure
t2 = linspace(-1,1);
x2=-2*t2;
y2=1 + 0*t2;
z2=pi/2 + t2;
plot3(x2,y2,z2)
hold off          % make sure you have this at the end
```

D.2.7 Exercises

1. Consider the vector function with parametric equations

$$x = \cos(t)\sqrt{4 - 0.25\cos^2(10t)} + 5,$$
$$y = \sin(t)\sqrt{4 - 0.25\cos^2(10t)} + 9,$$
$$z = 0.5\cos(10t) - 8.$$

(a) Show that, for any t, the space curve from these parametric equations lies on a sphere by finding the equation of the sphere, showing all work on paper. Hint: using these equations for x, y, and z, can you get something in the form of $(x-h)^2 + (y-k)^2 + (z-l)^2 = r^2$? Write the equation of the sphere. What is the center and radius of the sphere?

(b) Within MATLAB, plot the parametric equations to make the space curve. Make sure your domain is defined nicely so you get the entire graph and it is not a jagged curve.

(c) Within MATLAB, create a second plot of the space curve and the sphere (using the `sphere` command) so they appear in the same figure. Make the space curve black and thicker than the default.

2. Consider the vector function with parametric equations

$$x = \cos(\pi t),$$
$$y = \sin(\pi t),$$

$$z = 3\sin(\pi t).$$

(a) Find the vector form of the equation of the tangent line to the curve at $t = 1$.
(b) Find the vector form of the equation of the tangent line to the curve at $t = 1/2$.
(c) Find the point of intersection of these two lines.
(d) Within MATLAB, plot the parametric equations, the tangent lines and the point. Make sure your domain is defined nicely so you get the entire graph and it is not a jagged space curve. Make the space curve in black, the first tangent line in blue, the second in red, and mark the point of intersection with a black x.

3. Consider the surface $f(x, y) = 4x^2\sqrt{y + 15} + 2xe^{1-y}\ln(x)$.
   (a) Find the equation of the tangent plane to when $x = 1$ and $y = 1$. Show your work on paper.
   (b) Within MATLAB, graph the surface $z = f(x, y)$ and the tangent plane, marking the point $(1, 1, f(1, 1))$. Make the surface yellow, the tangent plane orange, and the point black. Make the domain for x to be from 0.000001 to 3 and the domain for y from 0 to 3.
   BONUS: Use the `AnimateView2` function found on the text website to create an animated gif picture.
   (c) Notice that the above question could have been phrased "find the linear approximation $L(x, y)$ of the function $f(x, y) = \ldots$ at ..." and the answer would have been the same.
   Use your answer above to approximate $f(1.01, 0.99)$ by calculating this approximation within MATLAB. Name the approximation calculation `fApprox` within MATLAB.
   (d) Have MATLAB calculate the actual value of $f(1.01, 0.99)$ (answer given to four decimal places by default). Name this calculation of f `fActual` within MATLAB.

D.3. Applications of double integrals

D.3.1 Calculating integrals and viewing regions

For these exercises, we will use the Symbolic Math Toolbox within MATLAB to calculate integrals.

For example, to calculate $\int_{-2}^{2}\int_{0}^{1}(x^2 + y^2)\,dy\,dx$ we can enter

```
>> syms x y
>> int1 = int(x^2 + y^2, y, 0, 1)
int1 =
```

```
x^2 + 1/3
>> int2 = int(int1, x, -2, 2)
int2 =
20/3
```

Consider the tetrahedron bounded by $x = 0$, $y = 0$, $z = 0$, and $z = 10 - 4x - 2y$. The following code allows the solid to be viewed in MATLAB.

First, we can see view the projection of the solid in the xy-plane.

```
x=linspace(0,2.5);
y=5-2*x;
fill([x,0],[y,0], [0.75, 0.75, 0.75])
axis([0,2.5,0,5.5])  % adjust the axis([xmin, xmax, ymin, ymax]) if necessary
title('Projection of Tetrahedron onto xy-plane')
```

The following code allows us to view the solid.

```
xdomain = linspace(0,2.5,81);          % adjust if necessary
ydomain = linspace(0,5,81);          % adjust if necessary
[x,y]=meshgrid(xdomain, ydomain);
z=10-4*x-2*y;
%%%
% BONUS IF CAN MODIFY BELOW APPROPRIATELY TO GET YOUR CORRECT 3-D REGION
% IF NOT, DELETE THIS SECTION
L=size(z);
for m=1:L(1)
    for n=1:L(2)
        if (y(m,n) > (5-2*x(m,n)))
            y(m,n) = 5-2*x(m,n);
        end
    end
end
z=10-4*x-2*y;
% END OF BONUS SECTION
%%%
colormap('Summer')  % defines color
C=0*z - 100;         % color for tetrahedron
meshz(x,y,z,C)
hold on
mesh(x,y,0*x,'EdgeColor',loygray)
hold off
xlabel('x'),ylabel('y'),zlabel('z')
view(65,20)      % adjust if necessary
```

D.3.2 Exercises

Use MATLAB to calculate the integrals needed to answer the following questions. Show all work on paper any calculations needed to set them up. Sketch the regions involved.

1. Find the mass and center of mass of the lamina that occupies the region D bounded by the parabolas $y = \frac{1}{81}x^2$ and $x = 9y^2$ and has density function $\rho(x, y) = \sqrt{x}$. Plot the figure using MATLAB.
2. A lamina occupies the region inside the circle $x^2 + y^2 = 4y$ but outside the circle $x^2 + y^2 = 4$. Find the center of mass if the density at any point is inversely proportional to its distance from the origin. (Hint: it may be useful to convert to polar coordinates.)
3. Consider the function

$$f(x, y) = \begin{cases} \frac{4}{25}xy & 0 \leq x \leq 5,\ 0 \leq y \leq 1, \\ 0 & \text{otherwise.} \end{cases}$$

   (a) Verify the function $f(x, y)$ is a joint density function.
   (b) If X and Y are random variables whose joint density function is the function f above, find $P\left(X \geq \frac{5}{2}\right)$.
   (c) Find $P\left(X \geq \frac{5}{2},\ Y \leq \frac{1}{2}\right)$.
   (d) Find the expected values of X and Y.

4. Consider the mass of the solid tetrahedron bounded by the xy-plane, the yz-plane, the xz-plane, and the plane $x/6 + y/5 + z/30 = 1$.
   (a) Set up a triple integral to find the mass of the solid tetrahedron if the density function is given by $\rho(x, y, z) = x + y$. Write an iterated integral in this form to find the mass of the solid:

$$m = \iiint_R f(x, y, z)\, dV = \int_A^B \int_C^D \int_E^F \underline{\hspace{4cm}}\, dz\, dy\, dx$$

   specifying the integrand and limits of integration A, B, C, D, E, and F.
   (b) Use MATLAB to calculate the mass m.
   (c) Use MATLAB to calculate the center of mass (your answer should be a point).
   (d) BONUS: Use MATLAB to draw the projection of the solid onto the xy-plane.
   (e) BONUS: Use MATLAB to draw the 3D solid.

5. A family of surfaces $\rho = 1 + \frac{1}{5}\sin(m\theta)s\sin(n\phi)$, where m and n are positive integers are known as "bumpy spheres." They can be used to model tumors. We will work with the bumpy sphere

$$\rho = 1 + \frac{1}{5}\sin(7\theta)\sin(8\phi).$$

(a) Use MATLAB to graph the bumpy sphere using spherical coordinates and then converting to rectangular coordinates to get x, y, z defined. This can be done by setting up the proper domains for θ and ϕ using `meshgrid`, calculating ρ and then creating x, y, and z using the `sph2cart` command. Use `mesh`, making the `EdgeColor` the color of your choice.

(b) Write on paper the integral that would be used to calculate the volume of this bumpy sphere.

(c) Use MATLAB to calculate the exact value of the volume of the bumpy sphere.

(d) Write the answer MATLAB gives for the volume on paper, writing it in correct notation. For example, if MATLAB gives the answer as `pi*sqrt(3)/2`, write $\frac{\pi\sqrt{3}}{2}$.

6. Consider the integral $\iiint_E f(x, y, z)\,dV$ where E is the solid bounded by $z = 0, x = 0, z = y - 8x$ and $y = 24$.

(a) Use MATLAB to draw the 3D solid.

(b) It may be helpful to sketch on paper the projections of the solid onto the 2D coordinate planes to help change the order of the iterated integrals. You can get help for these sketches by adjusting the view of the 3D solid within MATLAB.

(c) Express the integral $\iiint_E f(x, y, z)\,dV$ as an iterated integral in six different ways, where E is the solid bounded by $z = 0, x = 0, z = y - 8x$ and $y = 24$.

(d) Choose one of the above six iterated integrals to have MATLAB compute if $f(x, y, z) = xyz$ using the Symbolic Math Toolbox. Which one did you choose and what is the answer?

References

[1] Stormy Attaway, MATLAB: A Practical Introduction to Programming and Problem Solving, third edition, Elsevier: Butterworth-Heinemann, Amsterdam, 2013.

[2] Michael Barnsley, Fractals Everywhere, second edition, Morgan Kaufmann, 2000.

[3] Niraj Chokshi, How Powerball manipulated the odds to create a $1.5 billion jackpot, Washington Post, January 13, 2016, https://www.washingtonpost.com/news/post-nation/wp/2016/01/13/how-powerball-manipulated-the-odds-to-make-a-1-5-billion-jackpot-happen/.

[4] Robert L. Devaney, The Chaos Game, http://math.bu.edu/DYSYS/chaos-game/node1.html, 1995.

[5] Educational Testing Service, GRE graduate record examinations guide to the use of scores, http://www.ets.org/gre/guide, 2019.

[6] Temple H. Fay, The butterfly curve, The American Mathematical Monthly 96 (5) (1989) 442–443, https://doi.org/10.2307/2325155.

[7] C.D. Fryar, Q. Gu, C.L. Ogden, K.M. Flegal, Anthropometric reference data for children and adults: United States, 2011–2014, National Center for Health Statistics, Vital Health Statistics 3 (39) (2016), https://www.cdc.gov/nchs/data/series/sr_03/sr03_039.pdf.

[8] Golden Gate Bridge Highway & Transportation District, FAQ, https://www.goldengate.org/faq/#59.

[9] Golden Gate Bridge Highway & Transportation District, School Projects, https://www.goldengate.org/bridge/history-research/educational-resources/school-projects/, 2006–2018.

[10] Lester S. Hill, Cryptography in an algebraic alphabet, The American Mathematical Monthly 36 (6) (Jun. - Jul. 1929) 306–312.

[11] Lester S. Hill, Concerning certain linear transformation apparatus of cryptography, The American Mathematical Monthly 38 (3) (Mar. 1931) 135–154.

[12] Guinness World Records Limited, Fastest Lacrosse Shot, https://www.guinnessworldrecords.com/world-records/fastest-lacrosse-shot?fb_comment_id=605553549549656_647775638660780, 2019.

[13] Desmond J. Higham, Nicholas J. Higham, MATLAB Guide, second edition, Society for Industrial and Applied Mathematics, Philadelphia, 2005.

[14] Ben Joffe, Functions 3D: examples, https://www.benjoffe.com/code/tools/functions3d/examples, 2019.

[15] Lee W. Johnson, R. Dean Riess, Jimmy T. Arnold, Introduction to Linear Algebra, fifth edition, Addison-Wesley, Boston, 2002.

[16] Benoit B. Mandelbrot, The Fractal Geometry of Nature, 1983.

[17] The MathWorks, Inc., MATLAB Documentation, https://www.mathworks.com/help/matlab/, 2019.

[18] Todd Neller, The Game of Pig, Gettysburg College Department of Computer Science, http://cs.gettysburg.edu/projects/pig/piggame.html. (Accessed 30 September 2019).

[19] David Nicholls, Fractal ferns, https://www.dcnicholls.com/byzantium/ferns/fractal.html, 1998.

[20] Robert Osserman, Mathematics of the Gateway Arch, Notices of the American Mathematical Society 57 (2) (February 2010) 220–229.

[21] Rapid Tables, CMYK to RGB color conversion, https://www.rapidtables.com/convert/color/cmyk-to-rgb.html.

[22] SPACE.com Staff, What Is the Distance Between Earth and Mars?, Future US, Inc., https://www.space.com/14729-spacekids-distance-earth-mars.html, March 1, 2012.

[23] E.M. Standish, Keplerian Elements for Approximate Positions of the Major Planets, Solar System Dynamics Group, JPL/Caltech, https://ssd.jpl.nasa.gov/txt/aprx_pos_planets.pdf.

[24] James Stewart, Single Variable Calculus: Early Transcendentals, eighth edition, Cengage Learning, Boston, MA, USA, 2016.

[25] James Stewart, Multivariable Calculus: Early Transcendentals, eighth edition, Cengage Learning, Boston, MA, USA, 2016.

[26] J.D. Watson, F.H. Crick, Molecular structure of nucleic acids; a structure for deoxyribose nucleic acid, Nature 171 (4356) (April 1953) 737–738.

[27] Gareth Williams, Linear Algebra with Applications, seventh edition, Jones & Bartlett Publishing, 2009.

Index